Collins

Cambridge International
AS & A Level Further Mathematics

Further Mechanics

STUDENT'S BOOK

Anthony Alonzi, Chris Chisholm
Series Editor: Dr Adam Boddison

William Collins' dream of knowledge for all began with the publication of his first book in 1819.

A self-educated mill worker, he not only enriched millions of lives, but also founded a flourishing publishing house. Today, staying true to this spirit, Collins books are packed with inspiration, innovation and practical expertise. They place you at the centre of a world of possibility and give you exactly what you need to explore it.

Collins. Freedom to teach.

Published by Collins
An imprint of HarperCollins*Publishers*
The News Building
1 London Bridge Street
London
SE1 9GF

HarperCollins*Publishers* Macken House
39/40 Mayor Street Upper
Dublin 1
DO1 C9W8
Ireland

Browse the complete Collins catalogue at
www.collins.co.uk

British Library Cataloguing-in-Publication Data

A catalogue record for this publication is available from the British Library.

Commissioning editor: Jennifer Hall
In-house editor: Lara McMurray
Author: Anthony Alonzi/Chris Chisholm
Series editor: Dr Adam Boddison
Development editor: Tim Major
Project manager: Emily Hooton
Copyeditor: Penny Nicholson
Reviewer: Adele Searle
Proofreaders: Julie Bond/Joan Miller
Answer Checkers: David Hemsley/Just Content
Cover designer: Gordon MacGilp
Cover illustrator: Maria Herbert-Liew
Typesetter: Jouve India Private Ltd
Illustrators: Jouve India Private Ltd/Ken Vail Graphic Design
Production controller: Sarah Burke
Printed and bound by Ashford Colour Press Ltd

This book contains FSC™ certified paper and other controlled sources to ensure responsible forest management.

For more information visit: www.harpercollins.co.uk/green

Acknowledgements

The publishers wish to thank Cambridge Assessment International Education for permission to reproduce questions from past AS & A Level Mathematics and Further Mathematics papers. Cambridge Assessment International Education bears no responsibility for the example answers to questions taken from its past papers. These have been written by the authors. Exam-style questions and sample answers have been written by the authors.

The publishers wish to thank the following for permission to reproduce photographs. Every effort has been made to trace copyright holders and to obtain their permission for the use of copyright material. The publishers will gladly receive any information enabling them to rectify any error or omission at the first opportunity.

pvi sportpoint/Shutterstock, p1 sportpoint/Shutterstock, p23 Oleg Totskyi/Shutterstock, p75 Annette Shaff/Shutterstock, p103 ALEXANDER V EVSTAFYEV/Shutterstock, p129 PhilipYb Studio/Shutterstock, p147 giulia186/Shutterstock.

Full worked solutions for all exercises, exam-style questions and past paper questions in this book available to teachers by emailing international.schools@harpercollins.co.uk and stating the book title.

CONTENTS

Full worked solutions for all exercises, exam-style questions and past paper questions in this book available to teachers by emailing international.schools@harpercollins.co.uk and stating the book title.

INTRODUCTION

This book is part of a series of nine books designed to cover the content of the Cambridge International AS and A Level Mathematics and Further Mathematics syllabus. The chapters within each book have been written to mirror the syllabus, with a focus on exploring how the mathematics is relevant to a range of different careers or to further study. This theme of *Mathematics in life and work* runs throughout the series, with regular opportunities to deepen your knowledge through group discussion and exploring real-world contexts.

Within each chapter, examples are used to introduce important concepts and practice questions are provided to help you to achieve mastery. Developing skills in modelling, problem solving and mathematical communication can significantly strengthen overall mathematical ability. The practice questions in every chapter have been written with this in mind and include symbols to indicate which of these underlying skills are being developed. Exam-style questions are included at the end of each chapter and a bank of practice questions including real Cambridge past exam questions is included at the end of the book.

A range of other features throughout the series will help to optimise your learning. These include:

> **key information boxes** – highlighting important learning points or key formulae

> **commentary boxes** – tackling potential misconceptions and strengthening understanding through probing questions

> **stop and think** – encouraging independent thinking and developing reflective practice.

Key mathematical terminology is listed at the beginning of each chapter and a glossary is provided at the end of each book. Similarly, a summary of key points and key formulae is provided at the end of each chapter. Where appropriate, alternative solutions are included within the worked solutions to encourage you to consider different approaches to solving problems.

This book will build on the modelling concepts introduced in the earlier mechanics syllabus component, so that you can solve a wider range of real-life physical situations. Much of the book is centred on modelling dynamic motion, such as throwing balls through the air, bouncing balls off walls and floors and spinning balls around in vertical or horizontal circles. Hooke's law will be introduced to help model situations that involve elastic strings or springs.

Mechanics is an area of mathematics that is directly applicable to a broad range of careers including architecture and the design of theme park rides. This book provides direct applications for a range of pure mathematics topics, including first order differential equations. In particular, the use of Cartesian equations to model projectile motion demonstrates how your knowledge of quadratic equations and calculus can help solve real-world physical problems.

Modelling in mechanics

When modelling mechanics problems it is common to use diagrams as a visual aid, and these may include a range of measures such as force, acceleration, velocity, speed and distance. The convention is to use different types of arrow to represent different measures, as shown below.

Force   ⟶

Acceleration   ⟶≫

Velocity / speed   ⟶>

Distance   ⟷>

Particular 'modelling words' have specific meanings in the context of mechanics problems. The table below includes some of the modelling words you can expect to encounter.

Modelling word	Assumption
Light	The object has no mass
Smooth	There is no friction
Rough	There is friction
Inextensible / inelastic	The object cannot be stretched or squashed
Uniform	The same throughout (for example, uniform velocity)
Particle	A single point representing an object
Rigid	The object cannot bend
String	A line with no thickness
Rod	A rigid straight line with no thickness

FEATURES TO HELP YOU LEARN

Mathematics in life and work

Each chapter starts with real-life applications of the mathematics you are learning in the chapter to a range of careers. This theme is picked up in group discussion activities throughout the chapter.

Learning objectives

A summary of the concepts, ideas and techniques that you will meet in the chapter.

Language of mathematics

Discover the key mathematical terminology you will meet in this chapter. Throughout the chapter, key words are written in bold. These words are defined in the glossary at the back of the book.

Prerequisite knowledge

See what mathematics you should know before you start the chapter, with some practice questions to check your understanding.

Explanations and examples

Each section begins with an explanation and one or more worked examples, with commentary where appropriate to help you follow. Some show alternative solutions in the example or accompanying commentary to get you thinking about different approaches to a problem.

---

1 MOTION OF A PROJECTILE

Mathematics in life and work

In this chapter, you will study the motion of a projectile. It is important that you can calculate the path that the projectile takes, including its maximum height and its range. This skill is required in many different careers and it is also important in the worlds of sport and leisure – for example:

> If you were working as a javelin coach, you would need to understand what angle of release maximises the horizontal distance travelled.

> If you were working as a swimming coach, you would need to understand the principles of projectile motion to maximise the distance travelled during a dive.

> If you were designing a children's game that involved firing plastic rockets, you would need to understand the maximum possible height and range so as to display this information on the box.

LEARNING OBJECTIVES

You will learn how to:

> model the motion of a projectile as a particle moving with constant acceleration and understand any limitations of the model

> use horizontal and vertical equations of motion to solve problems involving the motion of projectiles, including finding the magnitude and direction of the velocity at a given time or position, the range on a horizontal plane and the greatest height reached

> derive and use the Cartesian equation of the trajectory of a projectile, including problems in which the initial speed and/or angle of projection may be unknown.

LANGUAGE OF MATHEMATICS

Key words and phrases you will meet in this chapter:

> parabola, projectile, range, trajectory

PREREQUISITE KNOWLEDGE

You should already know how to:

> use appropriate formulae for motion with constant acceleration in straight lines (horizontal and vertical)

> resolve a vector into its horizontal and vertical components.

Example 4

Sasha is practising basketball in her garden. She releases the ball from a height of 2 m, at a velocity of 10 m s^{-1} and at an angle of 40° above the horizontal. On its descent, it lands in a net that is 3 m above the ground.

a   For how long was the ball in the air?

b   How far away from the base of the net was Sasha standing?

Solution

a   To find how long the ball was in the air, you need to find the times at which the ball is 3 m above the ground.

Colour-coded questions

Questions are colour-coded (green, blue and red) to show you how difficult they are. Exercises start with more accessible (green) questions and then progress through intermediate (blue) questions to more challenging (red) questions.

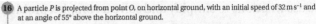

16 A particle P is projected from point O, on horizontal ground, with an initial speed of $32\,\mathrm{m\,s^{-1}}$ and at an angle of $55°$ above the horizontal ground.

  a Show that the greatest height of P is $34.4\,\mathrm{m}$, to 3 significant figures.

  b Find the amount of time that the height of P is between $\frac{1}{3}$ and $\frac{2}{3}$ of its maximum height.

17 A particle P is projected from a point O, on horizontal ground, with speed $12\,\mathrm{m\,s^{-1}}$ and at an angle of $30°$ above the horizontal. A second particle Q is also projected from O, with speed $V\,\mathrm{m\,s^{-1}}$ and at angle of $45°$ above the horizontal. P and Q move in the same vertical plane and the distance between P and Q when they strike the ground is $10\,\mathrm{m}$. Given that P lands closest to

Question-type indicators

The key concepts of problem solving, communication and mathematical modelling underpin your A level Mathematics course. You will meet them in your learning throughout this book and they underpin the exercises and exam-style questions. All Mathematics questions will include one or more of the key concepts in different combinations. We have labelled selected questions that are especially suited to developing one or more of these key skills with these icons:

(PS) Problem solving – mathematics is fundamentally about problem solving and representing systems and models in different ways. These include: algebra, geometrical techniques, calculus, mechanical models and statistical methods. This icon indicates questions designed to develop your problem-solving skills. You will need to think carefully about what knowledge, skills and techniques you need to apply to the problem to solve it efficiently.

These questions may require you to:

> use a multi-step strategy

> choose the most efficient method, or bring in mathematics from elsewhere in the curriculum

> look for anomalies in solutions

> generalise solutions to problems.

(C) Communication – communication of steps in mathematical proof and problem solving needs to be clear and structured and use algebra and mathematical notation so that others can follow your line of reasoning. This icon indicates questions designed to develop your mathematical communication skills. You will need to structure your solution clearly, to show your reasoning and you may be asked to justify your conclusions.

These questions may require you to:

> use mathematics to demonstrate a line of argument

> use mathematical notation in your solution

> follow mathematical conventions to present your solution clearly

> justify why you have reached a conclusion.

(MM) Mathematical modelling – a variety of mathematical content areas and techniques may be needed to turn a real-world situation into something that can be interpreted through mathematics. This icon indicates questions designed to develop your mathematical modelling skills. You will need to think carefully about what assumptions you need to make to model the problem and how you can interpret the results to give predictions and information about the real world.

These questions may require you to:

> construct a mathematical model of a real-life situation, using a variety of techniques and mathematical concepts

> use your model to make predictions about the behaviour of mathematical systems

> make assumptions to simplify and solve a complex problem.

Key information

These boxes highlight information that you need to pay attention to and learn, such as key formulae and learning points

KEY INFORMATION

The range of a projectile returning to the same height is given by $R = \dfrac{U^2 \sin 2\theta}{g}$ m.

Stop and think

Stop and think | How can you check that this is the maximum range?

These boxes present you with probing questions and problems to help you to reflect on what you have been learning. They challenge you to think more widely and deeply about the mathematical concepts, to tackle misconceptions and, in some cases, to generalise beyond the syllabus. They can be a starting point for class discussions or independent research. You will need to think carefully about the question and come up with your own solution.

Mathematics in life and work: Group discussions give you the chance to apply the skills you have learned to a model of a real-life mathematical problem involving a career that uses mathematics. Your focus is on applying and practising the concepts, and coming up with your own solutions, as you would in the workplace. These tasks can be used for class discussions, group work or as an independent challenge.

Summary of key points

At the end of each chapter, there is a summary of key formulae and learning points.

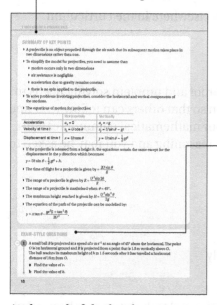

Exam-style questions

Practise what you have learned throughout the chapter with questions, written in examination style by our authors, that progress in order of difficulty.

The last **Mathematics in life and work** question draws together the skills that you have gained in this chapter and applies them to a simplified real-life scenario.

At the end of the book, test your mastery of what you have learned in the **Summary Review** section. Practise the basic skills and then go on some to try some carefully selected questions from Cambridge International A Level Mathematics questions and Further Mathematics past exam papers and exam-style questions. Extension questions, written by our authors, give you the opportunity to challenge yourself and prepare you for more advanced study.

1 MOTION OF A PROJECTILE

Mathematics in life and work

In this chapter, you will study the motion of a projectile. It is important that you can calculate the path that the projectile takes, including its maximum height and its range. This skill is required in many different careers and it is also important in the worlds of sport and leisure – for example:

> If you were working as a javelin coach, you would need to understand what angle of release maximises the horizontal distance travelled.

> If you were working as a swimming coach, you would need to understand the principles of projectile motion to maximise the distance travelled during a dive.

> If you were designing a children's game that involved firing plastic rockets, you would need to understand the maximum possible height and range so as to display this information on the box.

LEARNING OBJECTIVES

You will learn how to:

> model the motion of a projectile as a particle moving with constant acceleration and understand any limitations of the model

> use horizontal and vertical equations of motion to solve problems involving the motion of projectiles, including finding the magnitude and direction of the velocity at a given time or position, the range on a horizontal plane and the greatest height reached

> derive and use the Cartesian equation of the trajectory of a projectile, including problems in which the initial speed and/or angle of projection may be unknown.

LANGUAGE OF MATHEMATICS

Key words and phrases you will meet in this chapter:

> parabola, projectile, range, trajectory

PREREQUISITE KNOWLEDGE

You should already know how to:

> use appropriate formulae for motion with constant acceleration in straight lines (horizontal and vertical)

> resolve a vector into its horizontal and vertical components.

You should be able to complete the following questions correctly:

1 Find the horizontal and vertical components of these vectors.

a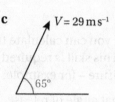

$F = 10\,\text{N}$

$23°$

b

$F = 12.6\,\text{N}$

$45°$

c

$V = 29\,\text{m s}^{-1}$

$65°$

d

$V = 0.2\,\text{m s}^{-1}$

$29°$

2 Use the appropriate equation of constant acceleration to answer these questions.

 a Find v when $u = 2\,\text{m s}^{-1}$, $a = 3\,\text{m s}^{-2}$ and $t = 4\,\text{s}$.

 b Find v when $u = 1\,\text{m s}^{-1}$, $a = 2\,\text{m s}^{-2}$ and $s = 4\,\text{m}$.

 c Find u when $v = 25\,\text{m s}^{-1}$, $a = 10\,\text{m s}^{-2}$ and $s = 16\,\text{m}$.

 d Find s when $u = 5\,\text{m s}^{-1}$, $a = 4\,\text{m s}^{-2}$ and $t = 5\,\text{s}$.

 e Find v when $u = 10\,\text{m s}^{-1}$, $t = 5\,\text{s}$ and $a = -2\,\text{m s}^{-2}$.

3 A ball is dropped from rest out of a window that is 10 m above the ground.

 a What is its velocity when it reaches the ground?

 b How long does it take to reach the ground?

4 A ball is thrown vertically upward from the ground, with initial speed $2\,\text{m s}^{-1}$.

 a After how long does it reach its maximum height?

 b What is its maximum height?

1.1 Projectile motion

Previously, you learnt how to model motion in a straight line horizontally and vertically. You found that vertical motion was affected by gravity but horizontal motion was not. In this section, you will look at motion that has both horizontal and vertical components.

In order to simplify the model and the calculations, you need to make the following assumptions:

 › motion occurs only in two dimensions

 › air resistance is negligible

 › acceleration due to gravity remains constant

 › there is no spin applied to the projectile.

Stop and think	Can you think of a real-life situation when acceleration due to gravity may not remain constant?

A particle that is launched into the air so that its subsequent motion is neither horizontal nor vertical is known as a **projectile**. Usually this means that the particle is launched at an angle, as shown in the diagram.

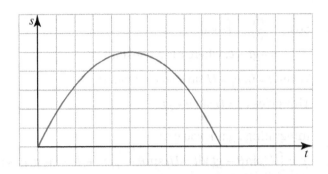

This graph of the path is known as the **trajectory** and it is a **parabola** (a quadratic graph). This is because horizontal motion is dependent only upon t while vertical motion is dependent upon t^2.

The motion is neither horizontal nor vertical, but can be considered as a combination of horizontal and vertical components.

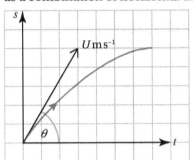 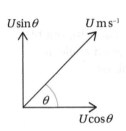

KEY INFORMATION

To solve problems involving projectiles, consider the horizontal and vertical components of the motion separately.

If the initial velocity is $U \, \mathrm{m \, s^{-1}}$ at an angle of θ above the horizontal, then the initial horizontal velocity is given by $U \cos \theta \, \mathrm{m \, s^{-1}}$ and the initial vertical velocity is given by $U \sin \theta \, \mathrm{m \, s^{-1}}$.

It is important to be precise when defining the angle. Here, you know exactly which angle is given because it is defined as being above the horizontal.

Horizontally, there is no acceleration. As a result, the horizontal velocity will be constant and you can say that $a_x = 0$ or $\dfrac{\mathrm{d}^2 x}{\mathrm{d}t^2} = 0$

since $a_x = \dfrac{\mathrm{d}^2 x}{\mathrm{d}t^2}$.

This means that:

$v_x = \dfrac{\mathrm{d}x}{\mathrm{d}t} = \int 0 \, \mathrm{d}t = c$ where c is a constant.

When $t = 0$, $v_x = U \cos \theta$ so $c = U \cos \theta$.

This gives $v_x = U \cos \theta$.

Integrating again, you get:

$x = \int (U \cos \theta) \, \mathrm{d}t = Ut \cos \theta + d$ where d is a constant.

In this case, when $t = 0$, $x = 0$ so $d = 0$.

This gives $x = Ut \cos \theta$.

Vertically, the magnitude of the acceleration is given by g.

This means that you can write $a_y = \dfrac{\mathrm{d}^2 y}{\mathrm{d}t^2} = -g$

The value of g is often given as $10 \, \mathrm{m \, s^{-2}}$ and this value should be used in all solutions unless you are instructed otherwise.

So $v_y = \dfrac{dy}{dt} = \int -g\,dt$

$\quad = -gt + c$, where c is a constant.

When $t = 0$, $v_y = U\sin\theta$ so $c = U\sin\theta$.

This gives $v_y = U\sin\theta - gt$.

> Notice the similarity between this and the equation of constant acceleration $v = u + at$.

Integrating again, you get:

$y = \int(U\sin\theta - gt)dt = Ut\sin\theta - \dfrac{1}{2}gt^2 + d$ where d is a constant.

In this case, when $t = 0$, $y = 0$ so $d = 0$.

This gives $y = Ut\sin\theta - \dfrac{1}{2}gt^2$.

> Notice the similarity between this and the equation of constant acceleration.

Stop and think   In the calculations above, why is $-g$ used instead of g?

The equations of motion for horizontal and vertical components of motion at time, t, with an initial velocity of U at an angle of θ above the horizontal can be summarised as follows.

	Horizontally	Vertically
Acceleration	$a_x = 0$	$a_y = -g$
Velocity at time t	$v_x = U\cos\theta$	$v_y = U\sin\theta - gt$
Displacement at time t	$x = Ut\cos\theta$	$y = Ut\sin\theta - \dfrac{1}{2}gt^2$

Stop and think   Working with a partner, show how these equations can be derived using the equations of uniformly accelerated motion.

If you know the initial velocity and the angle at which the projectile is released, you can use this information to find the maximum height and the **range** (the horizontal distance travelled) of the projectile.

Example 1

As part of a science experiment, a ball is projected into the air at $20\,\text{m s}^{-1}$ at an angle of $40°$ above the horizontal.

a How long does it take for the ball to reach its maximum height?

b What is the maximum height of the ball?

c What is the range of the ball?

Solution

$U = 20\,\text{m s}^{-1}$
$40°$

Most people find it useful to draw a diagram showing the key information from the question.

a The maximum height will occur when the vertical velocity is $0\,\text{m s}^{-1}$.

$U = 20\,\text{m s}^{-1}$, $\theta = 40°$, $v_y = 0\,\text{m s}^{-1}$, $g = 10\,\text{m s}^{-2}$

Substitute into $v_x = U \sin\theta - gt$:

$0 = 20 \sin 40° - 10t$

$t = 1.29\,\text{s}$

The time taken to reach the maximum height is 1.29 seconds.

If your final answer is not exact, write it to three significant figures or one decimal place if it is an angle, unless you're told otherwise. However, greater accuracy (at least 4 decimal places) should be retained for intermediate working.

b The maximum height will occur when $t = 1.286\,\text{s}$.

$U = 20\,\text{m s}^{-1}$, $\theta = 40°$, $g = 10\,\text{m s}^{-2}$, $t = 1.286\,\text{s}$

Substitute into $y = Ut \sin\theta - \frac{1}{2}gt^2$:

$y = 20 \times 1.286 \times \sin 40° - \frac{1}{2} \times 10 \times 1.29^2 = 8.26\,\text{m}$

The maximum height of the ball is 8.26 m.

c To find the range you need to find the time when the vertical displacement is 0 m.

$U = 20\,\text{m s}^{-1}$, $\theta = 40°$, $a_y = 10\,\text{m s}^{-2}$, $y = 0$

Substitute into $y = Ut \sin\theta - \frac{1}{2}gt^2$:

$0 = 20t \sin 40° - \frac{1}{2} \times 10 \times t^2$

Simplify:

$5t^2 - 20t \sin 40° = 0$

$t^2 - 4t \sin 40° = 0$

Factorise:

$t(t - 4 \sin 40°) = 0$

$t = 0$ or $t - 4 \sin 40° = 0$

So $t = 0\,\text{s}$ or $t = 4 \sin 40°\,\text{s}$

$t = 0\,\text{s}$ is when the ball left the ground.

This means the ball lands after $4 \sin 40°$ seconds.

You can now find the range by using this value for t.

$U = 20\,\text{m s}^{-1}$, $\theta = 40°$, $t = 4 \sin 40°\,\text{s}$

Substitute into $x = Ut \cos \theta$:

$x = 20 \cos 40° \times 4 \sin 40°$

$x = 39.4 \, \text{m}$

The range of the ball is 39.4 m.

Stop and think — Why is the vertical velocity $0 \, \text{m s}^{-1}$ at the maximum height?

At any instant, the direction in which the particle is travelling can be found. Trigonometry can be applied to the horizontal and vertical components of the particle's velocity.

Example 2

A batter hits a baseball at $53.9 \, \text{m s}^{-1}$, at an angle of $60°$ above the horizontal. It is assumed that the baseball is hit from ground level and moves freely under gravity. Find:

a the horizontal distance travelled by the baseball after 6 s

b the vertical displacement of the baseball after 6 s

c the speed of the baseball after 6 s

d the direction in which the baseball is travelling after 6 s

e the times at which the baseball is exactly 36.75 m above the ground.

Solution

a To find the horizontal displacement after 6 seconds, substitute $t = 6 \, \text{s}$ into the formula for horizontal displacement.

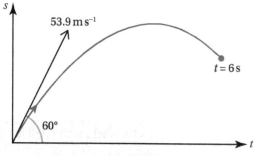

$U = 53.9 \, \text{m s}^{-1}$, $\theta = 60°$, $t = 6 \, \text{s}$

Substitute into $x = Ut \cos \theta$:

$x = 53.9 \times 6 \times \cos 60° = 162 \, \text{m}$

The horizontal distance travelled by the baseball after 6 s is 162 m.

b To find the vertical displacement after 6 seconds, substitute $t = 6\,\text{s}$ into the formula for vertical displacement.

$U = 53.9\,\text{m s}^{-1}$, $\theta = 60°$, $g = 10\,\text{m s}^{-2}$, $t = 6\,\text{s}$

Substitute into $y = Ut \sin\theta - \frac{1}{2}gt^2$:

$y = 53.9 \times 6 \times \sin 60° - \frac{1}{2} \times 10 \times 6^2 = 100\,\text{m}$

The vertical displacement of the baseball after 6 s is 100 m.

c The speed of the baseball at any time during its flight can be found by applying Pythagoras' theorem to the horizontal and vertical components of the baseball's velocity.

$U = 53.9\,\text{m s}^{-1}$, $\theta = 60°$, $g = 10\,\text{m s}^{-2}$, $t = 6\,\text{s}$

Substitute into $v_x = U \cos\theta$:
$v_x = 53.9 \times \cos 60° = 26.95\,\text{m s}^{-1}$

Hence the horizontal component of the velocity is $27.0\,\text{m s}^{-1}$.

Substitute into $v_y = U \sin\theta - gt$:

$v = 53.9 \times \sin 60° - 10 \times 6 = -13.32\,\text{m s}^{-1}$

Hence the vertical component of the velocity is $-13.32\,\text{m s}^{-1}$.

The speed is the magnitude of the velocity.

Applying Pythagoras' theorem:

$\text{Speed} = \sqrt{26.95^2 + (-13.32)^2} = 30.1\,\text{m s}^{-1}$

The speed of the baseball after 6 s is $30.1\,\text{m s}^{-1}$.

> The vertical component of the velocity is negative, which means it is in a downwards direction.

d Similarly to speed, the direction can be found using the horizontal and vertical components of the baseball's velocity, but this time by applying trigonometry. In the diagram below, $13.32\,\text{m s}^{-1}$ is the opposite side and $26.95\,\text{m s}^{-1}$ is the adjacent side, so you use the tangent ratio.

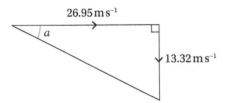

The angle will be $\tan^{-1}\left(\dfrac{13.32}{26.95}\right) = 26.3°$ below the horizontal.

e This question is similar to part **b** except that now you are given the vertical displacement and need to find possible values for t.

$U = 53.9\,\text{m s}^{-1}$, $\theta = 60°$, $g = 10\,\text{m s}^{-2}$, $y = 36.75\,\text{m}$

Substitute into $y = Ut\sin\theta - \frac{1}{2}gt^2$:

$36.75 = 53.9 \times t \times \sin 60° - \frac{1}{2} \times 10 \times t^2$

$36.75 = 53.9\sin 60° t - 5t^2$

$5t^2 - 53.9\sin 60° t + 36.75 = 0$

Solving using the quadratic formula:

$t = \dfrac{53.9\sin 60° \pm \sqrt{(53.9\sin 60°)^2 - 4 \times 5 \times 36.75}}{2 \times 5}$

$t = 0.9\,\text{s}$ and $8.5\,\text{s}$ (to 1 d.p.)

The times at which the baseball is exactly $36.75\,\text{m}$ above the ground are $0.9\,\text{s}$ and $8.5\,\text{s}$.

Mathematics in life and work: Group discussion

You are working as a swimming teacher and you have set up a diving competition to encourage children to learn to dive. The winner is the person that can travel the furthest distance from a stationary dive. Round one takes place in a deck-level swimming pool. (A deck-level pool is a pool where the water level is in line with the poolside.)

1 The first diver, Matilda, dives with a velocity of $2.5\,\text{m s}^{-1}$ at an angle of 40° above the water level. Find the distance from her starting point to where she enters the water.

2 The second diver, Alexander, enters the water $1.9\,\text{m}$ from his starting point. If he left at an angle of 44°, what was his initial velocity?

3 The third diver, Kay, leaves the poolside with a velocity of $3.2\,\text{m s}^{-1}$ and travels $1.8\,\text{m}$. At what angle did she leave the poolside?

4 When Matilda dived, it was found that the distance she travelled before entering the water was less than you calculated. Why may this be the case?

Exercise 1.1A

1 A tennis ball is thrown from ground level, with an initial speed of $25\,\text{m s}^{-1}$, at an angle of 45° above the horizontal.

   a Find the time the ball is in the air.

   b Calculate the range of the ball.

2 A cannonball is fired from the ground, with an initial speed of $40\,\text{m s}^{-1}$. It is fired at an angle of 45° above the ground.

   a Find the maximum height of the cannonball.

   b Find the range of the cannonball.

3 A tiger leaps from the ground, at an angle of 42° above the horizontal, at a speed of $8.8\,\text{m s}^{-1}$ in an attempt to get across a river that is 5 m wide.

   a For how long will the tiger be in the air?

   b Will the tiger reach the other side of the river? If so, by how far?

   c What assumptions have you made in the answers to parts **a** and **b**?

4 A baseball is hit from ground level, at a velocity of $30\,\text{m s}^{-1}$ and an angle of 53° above the horizontal. The baseball is modelled as a particle moving freely under gravity.

   a Find the horizontal distance travelled by the baseball after 3 s.

   b Find the vertical displacement travelled by the baseball after 3 s.

   c Find the speed of the baseball after 3 s.

5 A toy rocket is launched at a velocity of $22\,\text{m s}^{-1}$. If it is to hit a target 12 m away, at what angle should it be projected?

6 A rocket is launched from the ground, at a velocity of $25.0\,\text{m s}^{-1}$ and at an angle of θ above the horizontal. The rocket is modelled as a particle moving freely under gravity. After 4 seconds, the rocket lands x m from where it was launched.

   a Show that $\sin\theta = \dfrac{4}{5}$.

   b Find the value of x.

7 An animal jumps with an initial speed of $10\,\text{m s}^{-1}$, at an angle of 30° above the horizontal.

   a What is its maximum height?

   b What is its range?

   c For how long is the animal more than 1 m above the ground?

1.2 Projectiles launched from a given height

In **Section 1.1** you looked at projectiles launched from the ground. However, a projectile may be released and land at different heights. If you are throwing a ball, it will be released at the height of your hand and may land on the ground. When the initial height is not zero, you can make a slight adjustment to the equations of motion.

The horizontal component of the displacement of the projectile is unaffected and you can still use the formula $x = Ut \cos \theta$.

The vertical component of the displacement is affected. If the projectile is released from a given height (h), then when $t = 0$, $y = h$. This means that you use the formula $y = Ut \sin \theta - \frac{1}{2}gt^2 + h$ to find the displacement in the vertical direction, as shown in Example 3.

> **KEY INFORMATION**
>
> If a projectile is released from a height h, the equations remain the same except for the displacement in the y direction which becomes
>
> $y = Ut \sin \theta - \frac{1}{2}gt^2 + h$.

Example 3

A stone is thrown from the top of a vertical cliff, which is 30 m high, at a speed of $15 \, \text{m s}^{-1}$ and at an angle of $32°$ above the horizontal.

a Find the time for which the stone is in the air.

b Find the distance between the base of the cliff and where the stone lands.

Solution

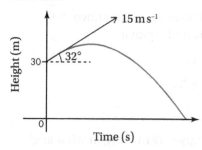

a The time of flight is found by finding when the vertical displacement, y, is 0 m.

$U = 15 \, \text{m s}^{-1}$, $\theta = 32°$, $g = 10 \, \text{m s}^{-2}$, $y = 0 \, \text{m}$, $h = 30 \, \text{m}$

Substitute into $y = Ut \sin \theta - \frac{1}{2}gt^2 + h$:

$0 = 15 \sin 32° \times t - \frac{1}{2} \times 10 \times t^2 + 30$

Simplify:

$5t^2 - 15 \sin 32° t - 30 = 0$

$t^2 - 3 \sin 32° t - 6 = 0$

Solve using the quadratic formula:

$t = \dfrac{3 \sin 32° \pm \sqrt{(-3 \sin 32°)^2 - 4 \times 1 \times (-6)}}{2 \times 1}$

$t = \dfrac{3 \sin 32° \pm \sqrt{9 \sin^2 32° + 24}}{2}$

$t = -1.780 \, \text{s}$ or $t = 3.370 \, \text{s}$

You only need the positive answer because time must be positive.

The time of flight for the stone is 3.37 seconds

Stop and think   What can you infer from the discarded negative solution?

b To find the distance from the base of the cliff to where the stone lands, you need to look at the horizontal displacement.

$U = 15\,\mathrm{m\,s^{-1}}$, $\theta = 32°$, $t = 3.370\,\mathrm{s}$

Substitute into $x = Ut\cos\theta$:

$x = 15 \times 3.370 \times \cos 32°$

$\quad = 42.87\,\mathrm{m}$

The stone lands 42.9 m from the base of the cliff.

In the next example, you will look at what happens when the projectile lands at a point higher than the starting point.

Example 4

Sasha is practising basketball in her garden. She releases the ball from a height of 2 m, at a velocity of $10\,\mathrm{m\,s^{-1}}$ and at an angle of 40° above the horizontal. On its descent, it lands in a net that is 3 m above the ground.

a For how long was the ball in the air?

b How far away from the base of the net was Sasha standing?

Solution

a To find how long the ball was in the air, you need to find the times at which the ball is 3 m above the ground. Since the ball entered the net on its descent, you need the larger of the two times. To find these times, consider the vertical motion.

$U = 10\,\mathrm{m\,s^{-1}}$, $\theta = 40°$, $g = 10\,\mathrm{m\,s^{-2}}$, $y = 3\,\mathrm{m}$, $h = 2\,\mathrm{m}$

Substitute into $y = Ut\sin\theta - \frac{1}{2}gt^2 + h$:

$3 = 10\sin 40° \times t - \frac{1}{2} \times 10t^2 + 2$

$5t^2 - 10\sin 40° t + 1 = 0$

Solve using the quadratic formula:

$t = \dfrac{10\sin 40° \pm \sqrt{(-10\sin 40°)^2 - 4 \times 5 \times 1}}{2 \times 5}$

$t = 0.1811\,\mathrm{s}$ and $t = 1.104\,\mathrm{s}$

The ball is released at a height of 2 m above the ground. On the ascent, it passes the height of 3 m after 0.1811 seconds. On the descent, it passes the height of 3 m after 1.104 s. So the time taken to reach the net was 1.1 seconds.

b To find the distance between Sasha and the net, you need to use the horizontal displacement.

$U = 10\,\text{m}\,\text{s}^{-1}$, $\theta = 40°$, $t = 1.104\,\text{s}$

Substitute into $x = Ut \cos \theta$:

$x = 10 \sin 40° \times 1.104$

$\quad = 7.10\,\text{m}$

Sasha is standing 7.10 m away from the net.

Mathematics in life and work: Group discussion

For the second round of the diving competition, the competitors dive from starting blocks that are 60 cm high.

1 The first diver, Matilda, leaves the starting block with a velocity of 2.5 m s^{-1} at an angle of 20° above the horizontal. How far away from the edge of the pool does she land?

2 Alexander enters the water 2.95 m from the edge of the pool. He left the block at an angle of 30° above the horizontal. What was his initial velocity?

3 Kay is determined to win this round. She knows she can leave the block with a velocity of 3.1 m s^{-1}. Between what two angles must she leave the starting block in order to win the competition?

4 You want to film the dives so that they can be analysed. You set your camera up in position but find that it will only film anything up to a height of 1.5 m. Will this allow you to record the full dives of all three swimmers?

Exercise 1.2A

1 A stone is thrown off the top of a cliff, with a velocity of 35 m s^{-1} and at an angle of 20° above the horizontal. The cliff is vertical and 45 m high.

   a Find the time for which the stone is in the air.

   b Find how far away from the base of the cliff the stone lands.

   c What assumptions have you made in this question?

2 A football rolls horizontally off a cliff, which is 20 m high, with a velocity of 20 m s^{-1}.

   a For how long is the ball in the air?

   b How far away from the cliff does the ball land?

3 A particle is launched from a point 49 m above the ground, at a velocity of 29.4 m s^{-1} and at an angle of 30° above the horizontal.

   a Find the time at which the particle hits the ground.

   b Find the horizontal distance travelled before the particle hits the ground.

   c Find the speed of the particle when it hits the ground.

   d Find the direction of the particle when it hits the ground.

4 A particle is launched from a point 50 m above the ground, at a velocity of 20 m s^{-1} and at an angle of 10° below the horizontal. Show that the horizontal distance travelled by the particle before it strikes the ground is about 56 m.

5 Andrei throws a stone horizontally from the top of a cliff at U m s^{-1}. The stone lands in the sea after 3 seconds. The stone is modelled as a particle moving freely under gravity.

  a Find the height of the cliff.

  b When the stone lands in the sea it is 73.5 m from the top of the cliff. Find the value of U.

6 A ball is kicked with velocity 25 m s^{-1}, at an angle of θ above the ground, from a point that is 4 m from a garage that is 2.8 m high. What is the smallest value of θ that will result in the ball landing on the roof of the garage?

1.3 Modelling the path of a projectile

If you start with the equations of motion for a projectile, you can derive specific formulae for calculating:

 » the time of flight

 » the range

 » the greatest height

 » the equation of the path for y in terms of x.

Time of flight for a particle returning to the same height

Consider a particle with an initial velocity of U m s^{-1}, launched at an angle of θ above the horizontal. Since the particle returns to the same height, you need to find t when $y = 0$ m.

Substitute into $y = Ut \sin \theta - \frac{1}{2}gt^2$:

$0 = Ut \sin \theta - \frac{1}{2}gt^2$

Rearrange:

$0 = t(U \sin \theta - \frac{1}{2}gt)$

So $t = 0$ or $U \sin \theta - \frac{1}{2}gt = 0$.

$t = 0$ s is when the particle in launched so $t = \dfrac{2U \sin \theta}{g}$ s.

The time taken for a particle to return to the same height is given by $t = \dfrac{2U \sin \theta}{g}$ s.

> **KEY INFORMATION**
>
> The time of flight for a projectile returning to the same height is given by
> $t = \dfrac{2U \sin \theta}{g}$.

Range

Consider the same particle returning to its original height.

You know that the time taken to return to the same height is

$t = \dfrac{2U \sin \theta}{g}$ and that the horizontal distance is given by $x = Ut \cos \theta$.

To find the range of the particle, substitute the time of the flight into the equation for the horizontal distance.

$$x = U \times \frac{2U\sin\theta}{g} \times \cos\theta.$$

$$x = \frac{2U^2\sin\theta\cos\theta}{g}.$$

$$x = \frac{U^2\sin 2\theta}{g}. \bullet \quad\text{Because } \sin 2\theta = 2\sin\theta\cos\theta.$$

The range of a particle returning to the same height is given by

$$R = \frac{U^2\sin 2\theta}{g} \text{ m}.$$

Angle of projection giving the maximum range

To find out the angle of projection that gives the maximum range, you need to find $\frac{dR}{d\theta}$.

If $R = \frac{U^2\sin 2\theta}{g}$, $\frac{dR}{d\theta} = \frac{2U^2\cos 2\theta}{g}$.

For a stationary point, $\frac{dR}{d\theta} = 0$.

$$\frac{2U^2\cos 2\theta}{g} = 0$$

$$\cos 2\theta = 0$$

$$2\theta = 90°$$

$$\theta = 45° \bullet \quad\text{Since } \theta \text{ is in the first quadrant.}$$

This means that the maximum range is obtained when the particle is projected at an angle of 45° above the horizontal. The maximum range is given by $R = \frac{U^2}{g}$.

> **KEY INFORMATION**
> The range of a projectile returning to the same height is given by $R = \frac{U^2\sin 2\theta}{g}$ m.

> **Stop and think** How can you check that this is the maximum range?

Greatest height

Consider the same particle, with an initial velocity of U m s^{-1}, launched at an angle of θ above the horizontal. The particle will reach its greatest height when its vertical velocity, v_y, is instantaneously zero. First, you need to know the time when this occurs.

$$v_y = U\sin\theta - gt$$

$$0 = U\sin\theta - gt$$

Rearrange to give $t = \frac{U\sin\theta}{g}$, which is the time taken to reach the maximum height.

Substitute this into $y = Ut \sin \theta - \frac{1}{2}gt^2$:

$$y = \frac{U^2 \sin \theta}{g} \sin \theta - \frac{1}{2}g\left(\frac{U \sin \theta}{g}\right)^2$$

$$y = \frac{U^2 \sin^2 \theta}{g} - \frac{U^2 \sin^2 \theta}{2g}$$

$$y = \frac{U^2 \sin^2 \theta}{2g}$$

The greatest height of a particle is given by $H = \frac{U^2 \sin^2 \theta}{2g}$ m.

> **KEY INFORMATION**
>
> The maximum height reached by a projectile is given by $H = \frac{U^2 \sin^2 \theta}{2g}$.

Equation of the path of a projectile

Consider the same particle, with an initial velocity of U m s^{-1}, launched at an angle of θ above the horizontal.

Assume that at time t s, the particle has a horizontal displacement of x m and a vertical displacement of y m.

Recall that time, t, is given by $\frac{x}{U \cos \theta}$.

Vertically, $u = U \sin \theta$ m s^{-1} and $a = -g$ m s^{-2}.

Substitute $t = \frac{x}{U \cos \theta}$ into $y = Ut \sin \theta - \frac{1}{2}gt^2$:

$$y = U \frac{x}{U \cos \theta} \sin \theta - \frac{1}{2}g\left(\frac{x}{U \cos \theta}\right)^2$$

$$y = x \tan \theta - \frac{gx^2}{2U^2 \cos^2 \theta}$$

$$y = x \tan \theta - \frac{gx^2 \sec^2 \theta}{2U^2}$$

$$y = x \tan \theta - \frac{gx^2(1 + \tan^2 \theta)}{2U^2}$$

The vertical displacement, y m, of the particle when the horizontal displacement is x m is given by $y = x \tan \theta - \frac{gx^2(1 + \tan^2 \theta)}{2U^2}$ m.

To derive the above equation, you started with the formulae $x = Ut \cos \theta$ and $y = Ut \sin \theta - \frac{1}{2}gt^2$ and eliminated the variable t to get y in terms of x.

> **KEY INFORMATION**
>
> The equation of the path of the projectile can be modelled by
>
> $$y = x \tan \theta - \frac{gx^2\left(1 + \tan^2 \theta\right)}{2U^2}.$$

Example 5

Two stones are launched from the same place at ground level and land at the same point on the ground. The first stone is launched at an angle of 40° to the horizontal, with a speed of 28 m s^{-1}. The second stone is launched at an angle of 55° to the horizontal. Find the greatest height attained by the second stone.

Solution

Although you could use the equations of motion for a projectile but it is quicker to use the formulae for range and greatest height.

The formula for range is given by $x = \dfrac{U^2 \sin 2\theta}{g}$.

For the first stone, with a speed of $28\,\mathrm{m\,s^{-1}}$ at an angle of $40°$,

the range is $\dfrac{28^2 \sin 80°}{10}$.

For the second stone, with a speed of $U\,\mathrm{m\,s^{-1}}$ at an angle of

$55°$, the range is $\dfrac{U^2 \sin 110°}{10}$.

Since these are equal, $\dfrac{U^2 \sin 110°}{10} = \dfrac{28^2 \sin 80°}{10}$.

Therefore $U^2 = \dfrac{28^2 \sin 80°}{\sin 110}$.

The formula for greatest height is given by $H = \dfrac{U^2 \sin^2 \theta}{2g}$.

For the second stone, the greatest height is

$\dfrac{28^2 \sin 80°}{\sin 110°} \times \dfrac{\sin^2 55°}{2g} = 27.6\,\mathrm{m}$.

Mathematics in life and work: Group discussion

You carry out some analysis on the three competitors in the diving competition and calculate that the maximum velocities at which Matilda, Alexander and Kay can leave the starting block are $5.2\,\mathrm{m\,s^{-1}}$, $4.9\,\mathrm{m\,s^{-1}}$ and $5.5\,\mathrm{m\,s^{-1}}$, respectively.

1 What are the furthest distances from the poolside that they could reach?

2 Work out out the equation of the trajectory of each diver, then plot all three graphs on the same axes. Check that the graphs confirm your answers to question 1.

3 Assuming that Matilda dives at an angle of $45°$ above the horizontal, what is the difference in the maximum range between her diving from the poolside and diving from the starting block?

4 What are Alexander's speed and direction 0.2 seconds after leaving the diving block?

5 Another coach tells you that the best angle at which to dive off the block is less than $45°$. Why do you think this is?

Exercise 1.3A

Ⓒ 1 A golf ball is struck, at an angle of $30°$ to the horizontal, at $U\,\mathrm{m\,s^{-1}}$ from ground level and moves freely under gravity.

    a Show that the golf ball returns to the ground after $\dfrac{U}{g}$ s.

    b Show that the golf ball travels $\dfrac{U^2\sqrt{3}}{2g}$ m horizontally before it returns to the ground.

    c Show that the greatest height achieved by the golf ball is $\dfrac{U^2}{8g}$ m.

C **2** The range of a particle is given by $x = \dfrac{U^2 \sin 2\theta}{g}$ m, where θ is the angle of projection in degrees and U is the velocity of projection in m s^{-1}. Two students, Mei and Xing, were asked to find the maximum range in terms of U, stating the angle for which the maximum range occurs.

Mei differentiated the function with respect to θ and put it equal to zero:

$$\frac{dx}{d\theta} = \frac{2U^2 \cos 2\theta}{g} = 0$$

$$\cos 2\theta = 0$$

$$2\theta = 270°$$

$$\theta = 135°$$

Hence the maximum range was $\dfrac{U^2 \sin 270}{g} = -\dfrac{U^2}{g}$ m.

Xing considered the graph of $y = \sin\theta$, which has a maximum value in the interval $0 \leqslant \theta < 360°$ of 1 when $\theta = 90°$, from which he deduced that the angle was given by $2\theta = 180°$.

Hence the maximum range was $\dfrac{U^2 \sin 180}{g} = 0$ m.

Neither Mei nor Xing has worked out the correct expression for the maximum range.

Explain what is incorrect for each set of working and hence find the maximum range in terms of U, stating the angle for which the maximum range occurs.

MM **3** A cannonball is fired from the ground, at 40 m s^{-1} at an angle of θ to the horizontal, and lands on the ground 60 m away after t s. The ground is assumed to be flat and horizontal.

a Show that $t = \dfrac{3}{2} \sec\theta$.

b By considering vertical motion, show that $80 \sin\theta = gt$.

c Hence, or otherwise, prove that $\sin 2\theta = \dfrac{3g}{80}$.

MM **4** A particle is projected at $U \text{ m s}^{-1}$ at an angle of $\arctan 2$ to the horizontal. After t s its vertical displacement is given by y m and its horizontal displacement is given by x m.

Show that, after t s, $y = \dfrac{x(4U^2 - 5gx)}{2U^2}$.

MM **5** A particle is projected at an angle of $\arcsin\dfrac{3}{5}$ above the horizontal at $V \text{ m s}^{-1}$. It returns to the same height after t s.

a Show that after t s the horizontal distance travelled by the particle is given by $\dfrac{24V^2}{25g}$ m.

b Prove that the horizontal range of a particle projected at an angle of α above the horizontal is the same as if the particle were to be projected at an angle of $(90° - \alpha)$ above the horizontal.

PS **6** Two particles, A and B, are launched at the same time in the same vertical plane, with A 60 m vertically above B. Particle A is launched horizontally and particle B at 45° above the horizontal. The particles are launched at speeds of 14 m s^{-1} and 28 m s^{-1}, respectively.

Show that the particles collide after travelling 39.8 m horizontally.

SUMMARY OF KEY POINTS

> A projectile is an object propelled through the air such that its subsequent motion takes place in two dimensions rather than one.

> To simplify the model for projectiles, you need to assume that:

> > motion occurs only in two dimensions

> > air resistance is negligible

> > acceleration due to gravity remains constant

> > there is no spin applied to the projectile.

> To solve problems involving projectiles, consider the horizontal and vertical components of the motions.

> The equations of motion for projectiles:

	Horizontally	Vertically
Acceleration	$a_x = 0$	$a_y = -g$
Velocity at time t	$v_x = U\cos\theta$	$v_y = U\sin\theta - gt$
Displacement at time t	$x = Ut\cos\theta$	$y = Ut\sin\theta - \frac{1}{2}gt^2$

> If the projectile is released from a height h, the equations remain the same except for the displacement in the y direction which becomes:

$$y = Ut\sin\theta - \frac{1}{2}gt^2 + h.$$

> The time of flight for a projectile is given by $= \dfrac{2U\sin\theta}{g}$.

> The range of a projectile is given by $R = \dfrac{U^2\sin 2\theta}{g}$.

> The range of a projectile is maximised when $\theta = 45°$.

> The maximum height reached is given by $H = \dfrac{U^2\sin^2\theta}{2g}$.

> The equation of the path of the projectile can be modelled by:

$$y = x\tan\theta - \frac{gx^2(1 + \tan^2\theta)}{2U^2}.$$

EXAM-STYLE QUESTIONS

1 A small ball B is projected at a speed of $v\,\text{m s}^{-1}$ at an angle of 45° above the horizontal. The point O is on horizontal ground and B is projected from a point that is 1.5 m vertically above O. The ball reaches its maximum height of h m 1.5 seconds after it has travelled a horizontal distance of 10 m from O.

   a Find the value of v.

   b Find the value of h.

2 A particle P is projected at an angle of 35° above horizontal ground. The particle returns to the horizontal ground and its range is 25 m.

   a Find the speed at which the particle is projected.

   b Find the velocity of the particle after 1 second after projection.

3 A small ball B is projected at a speed of 23 m s^{-1} at an angle of θ above the horizontal. The point O is on horizontal ground and B is projected from a point that is 20 m vertically above O. After t seconds, the ball reaches the horizontal ground at a horizontal distance of 10 m from O.

   a Find the value of θ.

   b Find the value of t.

4 A particle P is projected, from a height h above the ground, with speed 18 m s^{-1} and at an angle of 20° above the horizontal. The particle reaches ground level 3 seconds after projection.

   a Find the angle that the particle's velocity makes with the horizontal when it strikes the ground.

   b Find the speed of the particle when it strikes the ground.

   c Find the value of h.

5 A small ball B is projected, from a point O on horizontal ground, with speed 35 m s^{-1} and at an angle of 38° above the horizontal. On its downward trajectory, B strikes a raised horizontal platform that is 2 m above the height of the ground.

   a Find the time taken for B to strike the raised platform.

   b Find the speed at which B strikes the raised platform.

6 A stone is projected at a speed of 10 m s^{-1}, at an angle of 30° above the horizontal. The point O is on horizontal ground and the stone is projected from a point that is 19 m vertically above O. The stone strikes the ground at a horizontal distance of d m from O.

   a Find the velocity of the stone when it strikes the ground.

   b Find the value of d.

7 A small ball B is projected into the air, with an initial speed of u m s^{-1} and at an angle of θ above the horizontal ground. Having reached its maximum height of 15 m, B returns to the horizontal ground with a range of 15 m.

   a Find the value of θ.

   b Find the value of u.

8 A particle is projected into the air, at an angle of 39° to the horizontal ground and with an initial speed of 12 m s^{-1}.

   a Find the length of time for which the ball is above 1 m.

   b Find the speed and direction of the ball at the point when it is first 1 m above the ground.

9 Two particles, A and B, are both projected at the same angle to the horizontal and at the same speed. Initially, A is positioned at the point O on the horizontal ground and B is 25 m vertically above O. When the particles strike the ground, the horizontal distances of A and B from O are 60 m and 75 m, respectively.

a Find the angle of projection of the particles.

b Find the initial speed of the particles.

c 10 A particle is projected from O with an initial velocity of $5\,\text{m s}^{-1}$, at an angle of $30°$ above the horizontal. At time t s after projection the horizontal and vertically upward displacements of the particle from O are x m and y m, respectively.

a In the case where the particle is projected from the ground, express x and y in terms of t and show that the equation of the trajectory of the particle is $y = \dfrac{\sqrt{3}}{3}x - \dfrac{4}{15}x^2$.

b Given that the particle returns to the ground, find the range of the particle.

c In the case where the particle is projected from a platform that is 5 m above the ground, write down a new equation for the trajectory of the particle.

11 A stone is projected from a point O on horizontal ground. The equation of the trajectory of the stone is

$$y = 4x - 0.4x^2 + 1,$$

where x m and y m are, respectively, the horizontal and vertically upward displacements of the stone from O.

a Find the greatest height of the stone.

b Find the height above O at which the stone was projected.

c Find the range of the stone.

MM 12 A small ball B is projected from a point on horizontal ground with a speed $U\,\text{m s}^{-1}$ at angle of θ above the ground.

a Show that the range of B is $\dfrac{U^2 \sin 2\theta}{g}$.

b Find the maximum possible range of B.

13 In a competition, the aim is to throw a small ball so that it lands in a hole 2 m away from the starting point. Rajvir consistently throws the ball with an initial speed of $6.5\,\text{m s}^{-1}$.

a Given that the ball lands in the hole, find the two possible angles above the horizontal at which the ball can be projected.

b Rajvir decides to go for the smaller of the two angles. Calculate the time after projection when the ball reaches the hole.

PS **14** A particle is projected from a point O on horizontal ground. The equation of the trajectory of the stone is

$$y = x\tan\theta - \frac{gx^2\sec^2\theta}{40},$$

where x m and y m are, respectively, the horizontal and vertically upward displacements of the stone from O. Given that $y = 0.4$ when $x = 1$, find the two possible angles of projection.

MM **15** A particle P is projected with speed $20\,\text{m s}^{-1}$, at an angle of $60°$ above the horizontal, from a point O. At the same time, a particle Q is projected directly upwards at speed $V\,\text{m s}^{-1}$ from a point on the horizontal ground $15\,\text{m}$ away from O. P and Q move in the same vertical plane and they collide after they have been in motion for T s.

a Show that $T = 1.5\,\text{s}$.

b Find the value of V.

c Find the greatest height reached by P.

16 A particle P is projected from point O, on horizontal ground, with an initial speed of $32\,\text{m s}^{-1}$ and at an angle of $55°$ above the horizontal ground.

a Show that the greatest height of P is $34.4\,\text{m}$, to 3 significant figures.

b Find the amount of time that the height of P is between $\frac{1}{3}$ and $\frac{2}{3}$ of its maximum height.

PS **17** A particle P is projected from a point O, on horizontal ground, with speed $12\,\text{m s}^{-1}$ and at an angle of $30°$ above the horizontal. A second particle Q is also projected from O, with speed $V\,\text{m s}^{-1}$ and at angle of $45°$ above the horizontal. P and Q move in the same vertical plane and the distance between P and Q when they strike the ground is $10\,\text{m}$. Given that P lands closest to O, find the value of V.

18 A small ball B is projected from a point O, on horizontal ground, with speed $14\,\text{m s}^{-1}$ and at an angle of $35°$ above the horizontal. A tall vertical wall is $12\,\text{m}$ from O and perpendicular to the plane in which B moves. Given that the base of the window is $2\,\text{m}$ above the ground and the height of the window is $1.5\,\text{m}$, show that B passes through the window.

19 The equation of the trajectory of a small ball B projected from a fixed point O is

$$y = 5x - x^2,$$

where x and y are, respectively, the displacements, in metres, of B from O in the horizontal and vertically upward directions.

a Given that the direction of motion of B is at an angle of θ to the horizontal, show that

$$\tan\theta = 5 - 2x.$$

b Find the height of B when the direction of motion of B is $10°$ above the horizontal.

Mathematics in life and work

A swimmer dives with a speed of $5\,\mathrm{m\,s^{-1}}$ at an angle of $30°$ above the horizontal. He has the option of diving from a starting block (which is $60\,\mathrm{cm}$ high) or from the edge of the deck-level pool.

1   How much further will he dive if he uses the starting block, compared with diving from the poolside?

2   Which method of diving will allow him to enter the water at the smallest angle? You must show all your working out.

3   Following a training plan, the swimmer can now dive with an initial speed of $5.2\,\mathrm{m\,s^{-1}}$ at the same angle as before. How much further can he dive, using the starting block, if he follows the training plan?

2 EQUILIBRIUM OF A RIGID BODY

Mathematics in life and work

In this chapter, you will study the concept of moments and centre of mass. It is important that you are able to calculate moments around given points of an object and use this to help you to analyse situations involving systems in equilibrium. This skill is required in many different careers, especially those related to engineering – for example:

> If you were a structural engineer designing a viewing platform overlooking the edge of a canyon, you would need to understand moments to ensure that the platform will hold the weight of tourists.

> If you worked in construction and were responsible for operating a crane, you would need to understand moments and centre of mass to ensure that the weight you are trying to lift will not topple the crane.

> If you built statues you would need to have an understanding of centre of mass and moments to ensure that the statues would balance when outside and not be knocked over by strong winds.

LEARNING OBJECTIVES

You will learn how to:

> calculate the moment of a force about a point

> use the result that the effect of gravity on a right body is equivalent to a single force acting at the centre of mass of the body

> identify the position of the centre of mass of a uniform body, using considerations of symmetry

> use given information about the position of the centre of mass of a triangular lamina and other similar shapes

> determine the position of the centre of mass of a composite body by considering an equivalent system of particles

> use the principle that if a rigid body is in equilibrium under the action of coplanar forces, then the vector sum of the forces is zero and the sum of the moments of the forces about any point is zero, and the converse of this

> solve problems involving the equilibrium of a single rigid body under the action of coplanar forces, including those involving toppling or sliding.

LANGUAGE OF MATHEMATICS

Key words and phrases you will meet in this chapter:

> centre of mass, equilibrium, lamina, median, plane of symmetry, moment, sense, sliding, tilting, toppling

PREREQUISITE KNOWLEDGE

You should already know how to:

> use Pythagoras' theorem and trigonometry to find the magnitude and direction of a vector

> resolve a force into mutually perpendicular components

> use the formula $W = mg$ to convert between the mass and weight of an object

> apply the relationship $F \leqslant \mu R$ to problems involving friction and reaction (and $F = \mu R$ in the case of limiting equilibrium).

You should be able to complete the following questions correctly:

1 Find the weight, in N, of an apple with a mass of 125 g.

2 Find the horizontal and vertical components of a 9 N force acting at 52° to the horizontal.

3 Given a particle in limiting equilibrium with a friction force of $7.2g$ N and a reaction force of $18g$ N, find the coefficient of friction.

4 An object of mass 5 kg is pulled along a rough horizontal surface by a light piece of string inclined at 30° above the horizontal. The coefficient of friction between the surface and the object is 0.25. Given that the tension in the string is 20 N, find:

a the normal reaction force

b the friction force

c the acceleration of the object.

2.1 The moment of a force around a point

You will have already seen that forces can be used to accelerate objects along straight lines. However, as you will have seen when opening a door, sometimes forces will turn (rotate) objects instead. In the case of a door, this is due to hinges that attach the door to its frame.

When a force is applied to an object that is fixed or supported at a point along an axis that does not pass through the point, then that force will generate a turning force, known as a **moment**.

The formula for a moment is given by:

turning moment = force × perpendicular distance

Using SI units, the force is measured in newtons (N) and the distance in metres (m), giving the units N m for the moment. 1 N m is the same as 1 J (joule), the SI unit for energy.

For example, a force of 8 N at a perpendicular distance of 3 m would produce a moment of 24 N m.

Note that the formula confirms what you observe when you push a door. Pushing a door requires less effort, the further you push from the hinge. This is because, for the same moment, the force

> **KEY INFORMATION**
>
> moment
> = force × perpendicular
> distance
>
> Units are N m.

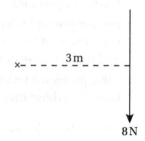

and distance are inversely proportional – as one increases, the other decreases.

When the distance is not perpendicular to the force, you need to resolve the force into mutually perpendicular components. In the diagram below, the horizontal component is $12\cos 50°$, which acts through the origin and so has a perpendicular distance of zero. The vertical component is $12\sin 50°$ and is the component perpendicular to the distance. In general, if the angle between the force and the distance is θ, then the moment is given by $Fd\sin\theta$, where F is the force and d is the distance. Note that the $Fd\cos\theta$ component acts through the turning point, so has no effect because $d = 0$.

KEY INFORMATION

For a force F acting at a distance d m at angle of θ, turning moment is given by $Fd\sin\theta$.

For the diagram, the moment is $12 \times 5\sin 50° = 46.0\,\text{N m}$.

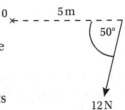

Depending upon the direction of the force, the moment may be clockwise or anticlockwise. This is known as the **sense** of the moment. By convention we refer to anticlockwise moments as positive and clockwise moments as negative.

The overall moment about a point can be found by adding the individual moments. For example, two clockwise forces of $8\,\text{N}$ at a perpendicular distance of $3\,\text{m}$ and $2\,\text{N}$ at a perpendicular distance of $5\,\text{m}$ have a combined moment of $24 + 10 = 34\,\text{N m}$ clockwise. However, if the first moment was clockwise and the second anticlockwise, then you would find the difference. The overall moment would be $24 - 10 = 14\,\text{N m}$ clockwise.

KEY INFORMATION

The sense of a moment is whether it is turning clockwise or anticlockwise. By convention, we refer to anticlockwise moments as positive and clockwise moments as negative.

Example 1

The diagram shows three forces acting about the origin.

Find the overall moment.

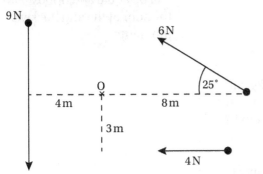

Solution

First, appreciate that the moment of the 4 N force is acting in a clockwise sense whereas the moments of the 9 N and 6 N forces are acting in an anticlockwise sense.

The perpendicular distance from the origin to the 9 N force is 4 m. The moment is $9 \times 4 = 36$ N m anticlockwise.

If the 4 N force is extended along its line of action, then it is easier to see that the perpendicular distance from the origin to the force is 3 m. The moment is $4 \times 3 = 12$ N m clockwise.

Make sure that the force and distance are perpendicular.

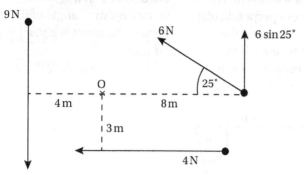

The 6 N force is acting at an angle of 25° to the horizontal, so the 8 m is not perpendicular to the force. If, however, the force is resolved into its horizontal and vertical components, then the vertical component ($6 \sin 25°$) is now perpendicular to the distance. The moment is $6 \sin 25° \times 8 = 20.29$ N m anticlockwise.

The overall moment is $36 + 20.29 - 12 = 44.3$ N m.

When calculating the overall moment, the anticlockwise moments are added and the clockwise moments are subtracted.

Since the answer is positive, the moment is 44.3 N m anticlockwise.

If the overall moment about a point is zero, then the individual clockwise and anticlockwise moments will balance. This means that the total of all the clockwise moments will have the same magnitude as the total of all the anticlockwise moments.

When an object is in **equilibrium**, the sum of the forces (and hence the resultant force) in any direction will be zero. The sum of the moments about any point will also be zero.

KEY INFORMATION

If the overall moment about a point is zero, then the sum of the clockwise moments will be equal and opposite to the sum of the anticlockwise moments.

Example 2

Nilesh, who weighs 54 kg, and Samvir, who weighs 42 kg, are playing on a seesaw. When Nilesh is 1.5 m from the centre, the seesaw is balanced. How far away from the centre is Samvir?

Solution

Since the seesaw is balanced, the clockwise moments must equal the anticlockwise moments.

It is always useful in this type of question to draw a diagram, marking clearly all of the forces.

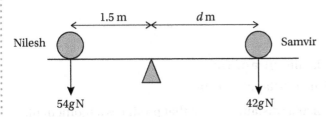

Nilesh's weight is acting in an anticlockwise direction and Samvir's in acting in a clockwise direction.

Moment caused by Nilesh = $1.5 \times 54g$

Moment created by Samvir = $-42g \times d$

This moment is negative because it is acting in a clockwise direction.

Since the seesaw is balanced, the sum of the moments is zero.

$$1.5 \times 54g - 42g \times d = 0$$

$$81 - 42d = 0$$

$$d = 1.93\,\text{m (to 3 s.f.)}$$

This means that Samvir is 1.93 m from the centre of the seesaw.

Mathematics in life and work: Group discussion

You are working for a company that creates models of characters from children's TV programmes. You have been asked to create some hanging decorations with models suspended from a light rod attached to a hook on the ceiling by a light string. The decorations need to balance correctly so that they remain horizontal.

1   Four characters of mass 30 g, 45 g, 60 g and 72 g must be attached to a light rod that is 1 m long. The string is attached to the centre of the rod. You attach the 72 g and 60 g objects 5 cm from each end of the rod. The 45 g object is 40 cm from the 72 g weight. Where should you position the 30 g object so the rod balances?

2   The second decoration has five characters of mass 21 g, 34 g, 42 g, 56 g and 70 g attached to a similar light rod. Find a way of placing the characters on the light rod so that it is balanced.

3   For the final decoration five characters of mass 20 g, 40 g, 60 g, 80 g and 100 g are to be positioned on a 1 m light rod so they are all 25 cm apart in ascending order (that is, the lightest one on the left and the heaviest one on the right). Where should the string be attached so that the rod balances?

4   Despite carefully calculating the location for each character to be attached in the final decoration, the rod is not balanced. Why do you think this may be?

Exercise 2.1A

1 **a** Find the moment, in Nm, of:

    i a force of 6 N acting at a perpendicular distance of 5 m

    ii a force of 3.5 N acting at a perpendicular distance of 12 m.

  b Find the force, in N, acting at a perpendicular distance of 4 m that produces a moment of:

    i 28 Nm   **ii** 15 Nm.

2 Forces P and Q act around a point O. Force P has a magnitude of 12 N and acts at a perpendicular distance of 4 m from O. The moment of P about O is clockwise. Force Q has a magnitude of 17 N and acts at a perpendicular distance of 3 m from O. The moment of Q about O is anticlockwise. Find the overall moment about O, stating its sense.

3 Points A, B, C and D lie on a straight line such that $AB = 4$ m, $BC = 5$ m and $CD = 6$ m. There is a 5 N force acting vertically upwards at A and a 3 N force acting vertically downwards at C.

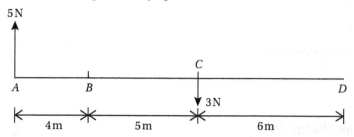

Find the moment about:

a A **b** B **c** C **d** D.

4 Find the moment about the origin O for each diagram.

a 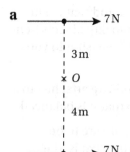   **b**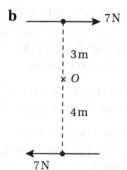

c 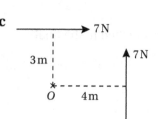   **d**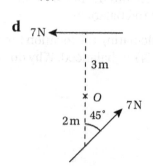

   Ⓒ Communication   ⓂⓂ Mathematical modelling   ⓅⓈ Problem solving

PS 5 Chen and Subash are trying to find the moment of a 9 N force about the origin O. The force acts through the point A at 30° to the horizontal, as shown in the diagram. The line OA is horizontal and 5 m in length.

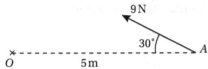

Chen resolves the force into horizontal and vertical components of $9\cos 30°$ N and $9\sin 30°$ N. She then multiplies $9\sin 30°$ N by 5 m, since $9\sin 30°$ N is the perpendicular component of the force. Her answer is 22.5 N m.

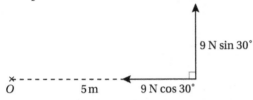

Subash sketches the force as a longer arrow and then marks the perpendicular distance between the origin and the force, which he labels as d. He then uses trigonometry to calculate that $d = 5\sin 30°$ m and multiplies this by 9 N. His answer is 22.5 N m.

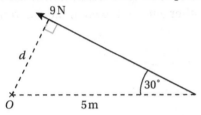

Are both methods valid? Whose method is more efficient?

PS 6 An 8 N and a 10 N force act through the same point, 3 m from the origin O. The 8 N force acts at 30° and the 10 N force acts at 60°, as shown in the diagram.

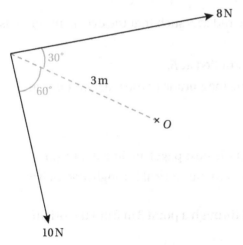

a Find the total moment about O.

b If the 10 N force acted at 50° instead, how would that affect the moment?

(PS) **7** Three forces act about a point X. Force A has a magnitude of 13 N and acts at a perpendicular distance of 4 m from the point, in a clockwise sense. Forces B and C both act in an anticlockwise sense. Force B has a magnitude of 6 N and acts at a perpendicular distance of 5 m. Force C has a magnitude of 11 N and acts at a perpendicular distance of d m.

a If the overall moment is zero, find the value of d.

b Find the overall moment if force C is increased to 15 N.

(MM) **8** A man M and a boy B sit on a seesaw that is 4 m long. The pivot of the seesaw is at the centre. The centre of mass of the seesaw is also at the centre, as shown in the diagram.

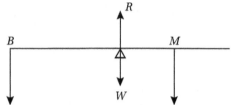

The boy sits at the end and has a mass of 36 kg. The man sits on the other side and has a mass of 60 kg. How far must the man sit from the centre of the seesaw to keep the seesaw balanced?

(MM) **9** A 10 m rod is supported by a pivot at point A. There is a 7 N force acting vertically downwards at one end and a 9 N force acting vertically upwards at the other end, as shown in the diagram. The rod is considered to be so light that it has no weight.

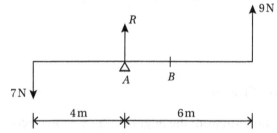

a Find the overall moment about the point A.

Point B is 2 m from A. A vertical force is applied to the rod at B such that the overall moment is now zero.

b i Find the magnitude and direction of the force applied at B.

ii By considering all the vertical forces, what is the magnitude of the reaction force at A such that the system is in equilibrium?

(C) **10** Samra was given the following problem.

The three forces T, U and V shown in the diagram (on the next page) are in equilibrium.

The magnitude of force T is q N and it acts at angle of α to the vertical through a point 8 m from the origin.

Force U acts at an angle of $(\alpha + 90°)$ to the horizontal through a point 3 m from the origin.

Force V acts through the origin.

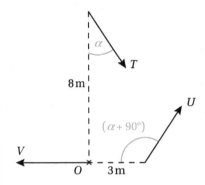

a Find the magnitude of U in terms of q.

b Find the value of α.

c Find the magnitude of V in terms of q.

Samra's solution is shown below.

a Resolving vertically:

$$T \cos \alpha = U \cos (90° + \alpha)$$
$$T \cos \alpha = U \cos 90° + U \cos \alpha$$
$$q \cos \alpha = 0 + U \cos \alpha$$
$$q \cos \alpha = U \cos \alpha$$
$$U = q \, \text{N}$$

b Taking moments about O:

$$U \times 3 \sin (\alpha + 90°) - q \times 8 \sin \alpha = 0$$
$$q \times 3 \sin (\alpha + 90°) - q \times 8 \sin \alpha = 0$$
$$3 \sin (\alpha + 90°) = 8 \sin \alpha$$
$$3 \sin \alpha + \sin 90° = 8 \sin \alpha$$
$$1 = 5 \sin \alpha$$
$$\sin \alpha = 0.2$$
$$\alpha = \sin^{-1} (0.2) = 11.5°$$

c Resolving horizontally:

$$V = T \sin \alpha + U \sin (90° + \alpha)$$
$$V = q \sin \alpha + q \sin (90° + \alpha)$$
$$V = q \sin \alpha + q \sin 90° + q \sin \alpha$$
$$V = q \sin \alpha + q \sin 90° + q \sin \alpha$$
$$V = 2q \sin 11.5° + q$$
$$V = 2q \times 0.2 + q$$
$$V = 1.4q \, \text{N}$$

Assess Samra's solution.

Where there are mistakes, provide the correct solution.

2.2 The centre of mass of a uniform body

The mass of a body is the quantity of matter that makes it up. Every part of an object forms part of its overall mass. When modelling an object as a particle, you can think of the mass of the object as being concentrated at one point. This point is known as the **centre of mass**. In this section, you will learn how to calculate the position of the centre of mass for a uniform body.

Suppose you have a system of particles as follows.

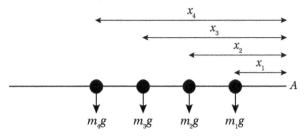

The total moment around $A = m_1gx_1 + m_2gx_2 + m_3gx_3 + m_4gx_4 + \ldots + m_ngx_n$

This whole system can be replaced by one particle of the same mass as all the others combined, acting at the centre of mass. You use the notation M to represent the total mass and $\bar{x}$ to represent the perpendicular distance between A and the centre of mass.

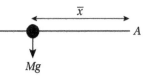

The total moment around $A = Mg\bar{x}$

Since these two systems are equal, the moments must be equal.

So $Mg\bar{x} = m_1gx_1 + m_2gx_2 + m_3gx_3 + m_4gx_4 + \ldots$

$M\bar{x} = m_1x_1 + m_2x_2 + m_3x_3 + m_4x_4 + \ldots$

$M\bar{x} = \sum_{i=1}^{i=n} m_ix_i$

This method allows you to calculate the centre of mass of uniform objects.

KEY INFORMATION

The centre of mass of a system of particles in a straight line can be found using

$M\bar{x} = \sum_{i=1}^{i=n} m_ix_i$

where M = the total mass of all particles and $\bar{x}$ is the perpendicular distance from the centre of mass to the point at which moments are being taken.

Example 3

For the following system of particles, which are all attached to a light rod, find the distance from A to the centre of mass.

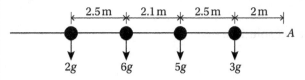

Solution

To find the centre of mass you need to use $M\bar{x} = \sum_{i=1}^{i=n} m_i x_i$

In this case $M = 3 + 5 + 6 + 2 = 16$

So $16\bar{x} = 3 \times 2 + 5 \times 4.5 + 6 \times 6.6 + 2 \times 9.1$

$16\bar{x} = 86.3$

$\bar{x} = 5.39\,\text{m}$ (to 3 s.f.)

The centre of mass of this system of particles is 5.39 m from A.

> The rod is light so you can ignore it in your calculation.

> You use the mass rather than the weight, so do not include g.

> Remember to use the distance from A rather than the distance between each particle.

Stop and think In calculations in this section (and in **Sections 2.3** and **2.4**), the density of the objects is ignored. Why is it acceptable to do this?

Exercise 2.2A

1 Find the position of the centre of mass for each system of particles.

a
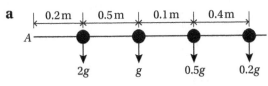

b
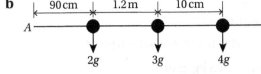

c
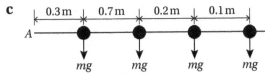

d
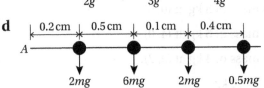

2 A light rod AB of length 2 m has particles, each of mass 2 kg, attached as shown below. The particles are equally spaced. Find the distance of the centre of mass from A.

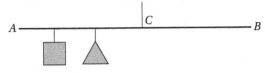

MM 3 A system of particles attached to a light rod has centre of mass 1.2 m from A. The first particle is 20 cm from A with a mass of m kg, the second particle is 1.3 m from A with a mass of $2m$ kg and the third particle is 2.5 m from A with a mass of 3.2 kg. Find m.

MM 4 A hanging decoration has shapes attached to a light plastic rod that is hung by a piece of string positioned at C, which is 40 cm from A. The square has a mass of 200 g and is 10 cm from A and the triangle has a mass of 100 g and is 30 cm from A. You need to attach a rectangle of mass 150g. Where would you position the rectangle to make point C the centre of mass?

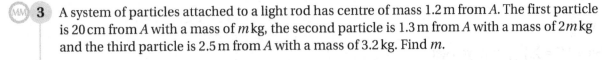

C 5 Three particles of mass m kg, $2m$ kg and 3 kg are positioned at points $(1, 0)$, $(7, 0)$ and $(9, 0)$, respectively. The centre of mass of these particles is at $(6, 0)$. Find m showing all of the key steps in your solution.

If you have a system of particles positioned in a plane instead of a straight line, you can extend the results to calculate the centre of mass.

If each particle has position (x_i, y_i) with mass m_i:

$M\bar{x} = \sum_{i=1}^{i=n} m_i x_i$

$M\bar{y} = \sum_{i=1}^{i=n} m_i y_i$

which gives the position of the centre of mass at $(\bar{x}, \bar{y})$.

Example 4

A light triangular framework has particles attached to each corner, as shown in the diagram below. Vertex A is at $(0, 0)$. Find the coordinates of the centre of mass.

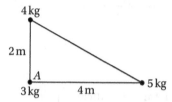

Solution

Point A is the origin, which gives:

a mass of 3 kg at $(0, 0)$

a mass of 5 kg at $(4, 0)$

a mass of 4 kg at $(0, 2)$.

$M = 12$ kg.

You can then find $\bar{x}$ and $\bar{y}$ separately.

Using $M\bar{x} = \sum_{i=1}^{n} m_i x_i$ you get:

$12\bar{x} = 3 \times 0 + 5 \times 4 + 4 \times 0$

$12\bar{x} = 20$

$\bar{x} = \frac{5}{3}$

Using $M\bar{y} = \sum_{i=1}^{n} m_i y_i$ you get:

$12\bar{y} = 3 \times 0 + 5 \times 0 + 4 \times 2$

$12\bar{y} = 8$

$\bar{y} = \frac{2}{3}$

So the coordinates of the centre of mass are $\left(\frac{5}{3}, \frac{2}{3}\right)$.

Exercise 2.2B

1 Find the coordinates of the centre of mass of each system of particles. Round answers to 3 significant figures where necessary.

a
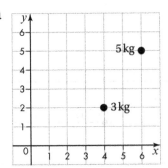

b
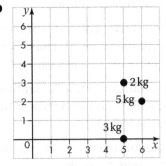

c
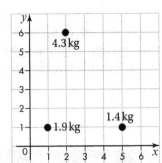

d
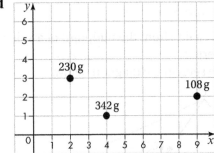

2 A light rectangular framework $ABCD$ has particles attached at each vertex. Vertex A is at $(0, 0)$. Find the coordinates of the centre of mass.

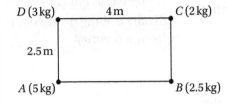

3 A light rectangular framework with vertices at $(2, 5)$, $(6, 5)$, $(2, 8)$ and $(6, 8)$ has stones of equal mass attached at each vertex. Find the coordinates of the centre of mass.

4 Masses are attached to each vertex of a light quadrilateral framework as follows:

$2\,\text{kg}$ at $(1, 1)$, $3\,\text{kg}$ at $(8, 1)$, $5\,\text{kg}$ at $(4, 5)$ and $m\,\text{kg}$ at $(2, 3)$.

If the centre of mass is at $(4, \bar{y})$, find m and $\bar{y}$.

5 The centre of mass of the following framework is at $(1.6, 2.5)$ with vertices at $A(0, 5)$, $B(4, 5)$, $C(4, 0)$ and $D(0, 0)$. Find m and n. Show your working clearly.

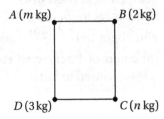

2.3 The centre of mass of a uniform lamina

In this section, you will look at techniques for finding the centre of mass of a uniform **lamina**. A lamina is a flat object that has negligible thickness. If its mass is spread evenly throughout its area, then it is called a uniform lamina.

If a uniform lamina has an axis of symmetry, then its centre of mass lies on this axis of symmetry. If it has more than one axis of symmetry, then its centre of mass lies where the axes intersect.

For example, in the uniform rectangular lamina shown here, the centre of mass will be where the two axes of symmetry intersect.

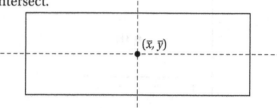

Using the same technique, the centre of mass of a circle of uniform mass is at its centre.

Triangular lamina

A **median** is a line from the midpoint of a side to the opposite vertex.
This can be seen in the diagram below.

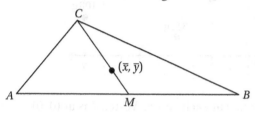

Let M be the midpoint of AB. The median is the line CM. The centre of mass, $(\bar{x}, \bar{y})$, is $\frac{2}{3}$ of the distance along the median from C.

You would get the same point if you used either of the other two medians.

Circular sector of radius r and angle 2α

For a sector of a circle with radius r, the centre of mass lies on the line that divides the sector into two equal smaller sectors. This means that if the angle at the centre of the sector is 2α, then you have an angle of α on either side of the line on which the centre of mass lies.

The centre of mass $(\bar{x}, \bar{y})$ of a circular sector of radius r and angle 2α is $\dfrac{2r\sin\alpha}{3\alpha}$ from the centre of the circle along this line of symmetry, where α is measured in radians.

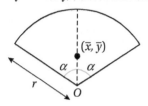

> The centres of mass of common shapes are given in the formula booklet.

> **KEY INFORMATION**
>
> The centre of mass of a triangular lamina is $\frac{2}{3}$ of the distance along the median from any vertex.

> **KEY INFORMATION**
>
> The centre of mass of a circular sector of radius r and angle 2α is $\dfrac{2r\sin\alpha}{3\alpha}$ from the centre of the circle, where α is measured in radians.

For a semicircle, $2\alpha = \pi$ so $\alpha = \frac{\pi}{2}$. This means that the centre of mass is $\frac{2r}{\frac{3\pi}{2}} = \frac{4r}{3\pi}$ from the centre.

Example 5

The base of a semicircular uniform lamina passes through the points $(2, 4)$ and $(6, 4)$. Find the coordinates, to 3 significant figures, of its centre of mass.

Solution

It is always useful to start with a diagram.

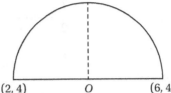

$(2, 4) \qquad O \qquad (6, 4)$

From the diagram, you can see that the centre of the circle is the point $(4, 4)$ and the radius is 2. By considering the lines of symmetry, you know that the centre of mass will lie somewhere on the line $x = 4$.

To find how far along this line it is, you need to find the value of $\frac{2r\sin\alpha}{3\alpha}$, where 2α is the angle of the sector in radians.

In this case $2\alpha = \pi$, so $\alpha = \frac{\pi}{2}$.

So $\dfrac{2r\sin\alpha}{3\alpha} = \dfrac{2 \times 2 \sin\frac{\pi}{2}}{3 \times \frac{\pi}{2}} = 0.849$.

This means that the centre of mass of this lamina is at the point $(4, 4.85)$.

> The centre of mass is 0.849 above the centre of the circle. Since the centre of the circle is at $(4, 4)$ you need to add 4 onto this value to get 4.85 (to 3 s.f.).

A lamina may be made up of a combination of shapes. The next two examples will explain how to deal with these types of lamina.

Example 6

Find the centre of mass of the lamina shown here.

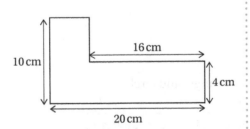

Solution

To find the centre of mass of this lamina you need to split it into two rectangles, A and B.

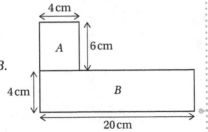

> This shape could have been split up into two different rectangles or a greater number of smaller shapes. The answer would be the same. The aim is to make the calculations that follow as simple as possible.

If you take the bottom left corner as the origin, you can make use of the lines of symmetry of each rectangle to work out that the centre of mass of A is at the point $(2, 7)$ and the centre of mass of B is at the point $(10, 2)$. Since this is a uniform lamina, the mass of each rectangle will be proportional to its area.

You can use a table to work out the centre of mass of the lamina.

Shape	Area	x-coordinate of centre of mass	y-coordinate of centre of mass
A	80	2	7
B	24	10	2
Lamina	104	$\bar{x}$	$\bar{y}$

Since the area is proportional to the mass:

$104\bar{x} = 80 \times 2 + 24 \times 10$

$104\bar{x} = 400$

$\quad \bar{x} = 3.85$ (to 3 s.f.)

$104\bar{y} = 80 \times 7 + 24 \times 2$

$104\bar{y} = 608$

$\quad \bar{y} = 5.85$ (to 3 s.f.)

So the centre of mass of this uniform lamina is at the point $(3.85, 5.85)$.

Example 7

Find the centre of mass of the shaded uniform lamina below. The circle has centre $(5, 4)$ and a radius of 1.5.

(3, 8)   (10, 8)

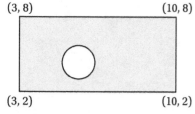

(3, 2)   (10, 2)

Solution

Using the lines of symmetry of the rectangle and circle, complete a table.

Shape	Area	x-coordinate of centre of mass	y-coordinate of centre of mass
Rectangle	42	6.5	5
Circle	$\dfrac{9}{4}\pi$	5	4
Lamina	$42 - \dfrac{9}{4}\pi$	$\bar{x}$	$\bar{y}$

Since the area is proportional to the mass:

$$(42 - \frac{9}{4}\pi)\bar{x} = 42 \times 6.5 - \frac{9}{4}\pi \times 5$$

$$\bar{x} = 6.80$$

$$(42 - \frac{9}{4}\pi)\,\bar{y} = 42 \times 5 - \frac{9}{4}\pi \times 4$$

$$\bar{y} = 5.20$$

Subtract $\frac{9}{4}\pi \times 5$ because the shape has been removed.

So the centre of mass of this lamina is at the point (6.80, 5.20) to 3 s.f.

Exercise 2.3A

1 Find the coordinates of the centre of mass of each lamina.

a
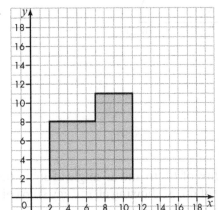

b
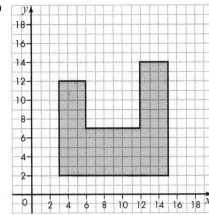

c
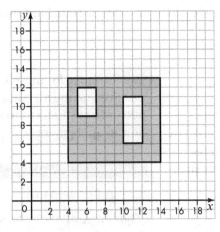

d
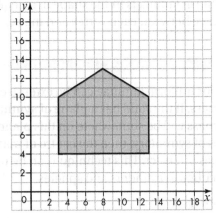

e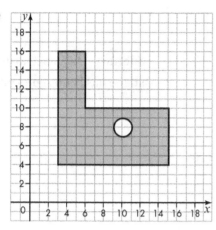

f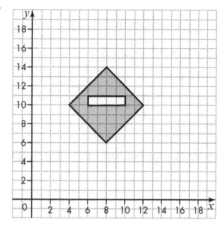

2 Find the coordinates of the centre of mass of this semicircular uniform lamina.

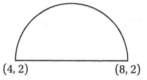

$(4, 2)$ $(8, 2)$

3 A uniform lamina is in the shape of a quarter of a circle with a radius of 4, as shown. Find the coordinates of its centre of mass.

$(4, 2)$

4 A uniform lamina is in the shape of a sector with a radius of 10 cm and an angle of 30° at the centre, O. Find the distance between O and the centre of mass, showing all the key steps of your solution.

5 A uniform lamina is in the shape of an equilateral triangle ABC with sides of 3 cm. Find the distance between each vertex and the centre of mass.

2.4 The centre of mass of a uniform solid

The rules for uniform laminae can be extended to solids of uniform mass. If the solid has a **plane of symmetry**, then the centre of mass will be somewhere on this plane. If it has more than two planes of symmetry, then the centre of mass will be where these planes intersect. This technique can be used on shapes such as cubes, cuboids, cylinders and spheres.

> The formulae for these shapes are given in the formula booklet but you need to know how to apply them.

Solid hemisphere

The centre of mass of a solid hemisphere with radius r is $\frac{3}{8}r$ from the centre of the hemisphere along the line connecting the centre of the base and the highest part of the hemisphere.

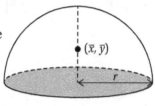

KEY INFORMATION

The centre of mass of a solid hemisphere with radius r is $\frac{3}{8}r$ from the centre of the hemisphere, on the line connecting the centre of the base and the highest part of the hemisphere.

Cone or pyramid

The centre of mass of a solid cone or right pyramid with height h is $\frac{3}{4}h$ below the vertex.

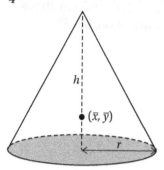 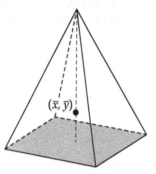

If you are able to work out the centre of mass of the solids above, you will be able to work out the centre of mass of composite solids made up of these shapes.

> **KEY INFORMATION**
>
> The centre of mass of a solid cone or right pyramid with height h is $\frac{3}{4}h$ below the vertex.

> This technique will only work if the density of each solid is the same.

Example 8

An ornament is made by placing a square-based pyramid of height 8 cm on top of a cube of side 10 cm, as shown below. Both solids are made from the same material. Calculate the position of the centre of mass of this shape.

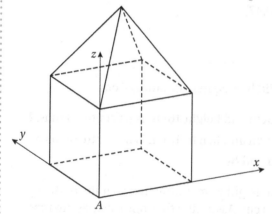

Solution

To help describe the location of points, it is helpful to add in the x, y and z axes. Point A above has coordinates $(0, 0, 0)$.

Both the cube and the pyramid have the same density so you can assume that the mass is uniformly spread throughout the combined solid.

The cube has sides of length 10 cm so, by looking at the symmetry, you can say the centre of mass is at the point $(5, 5, 5)$.

The pyramid has height 8 cm so its centre of mass will be $\frac{3}{4} \times 8 = 6$ cm below the vertex. By symmetry, the x and y coordinates will both be 5. Because the cube has height 10 cm, the centre of mass of the pyramid must be at the point (5, 5, 12).

> In this example, the centre of mass is 6 cm below the vertex, which means it is 2 cm above the centre of the base.

Shape	Volume	x-coordinate of centre of mass	y-coordinate of centre of mass	z-coordinate of centre of mass
Cube	1000	5	5	5
Pyramid	$\frac{800}{3}$	5	5	12
Solid	$\frac{3800}{3}$	$\bar{x}$	$\bar{y}$	$\bar{z}$

Since the x- and y-coordinates of the centre of mass are both 5 for the cube and the pyramid, the x- and y-coordinates of the solid must be also be 5.

> You can also deduce this from the symmetry of the solid.

Find the z-coordinate of the centre of mass:

$$\frac{3800}{3}\bar{z} = 1000 \times 5 + \frac{800}{3} \times 12$$

$$\bar{z} = 6.47 \text{ (to 3 s.f.)}$$

So the centre of mass of this solid is at the point (5, 5, 6.47).

Exercise 2.4A

1. Calculate the position of the centre of mass of a solid hemisphere of radius 4 cm.

2. **a** How far above the base of a solid cone of radius 6 cm and height 16 cm is its centre of mass?
   b How far above the base of a solid cone of radius 8 cm and height 16 cm is its centre of mass?
   c What can you infer from your answers to parts **a** and **b**?

3. A square-based pyramid with side of base 4 cm and height 6 cm is attached so it fits perfectly onto the end of a cuboid with sides 8 cm, 4 cm and 4 cm. Assuming that both shapes are made from the same material, find the position of the centre of mass of the new solid.

4. A solid hemisphere of radius 5 cm is attached to the base of an upside-down cone with radius 5 cm and height 10 cm. Assuming that the origin is at the vertex of the cone, find the coordinates of the centre of mass of the new solid.

5. A hemisphere is lying flat on a table. Its centre of mass is 6 cm above the top of the table. What is the radius of the hemisphere?

6. The cross-section of a solid step is shown in the diagram. The step is 50 cm long. Find the position of its centre of mass. Show your working clearly.

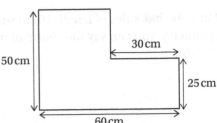

7 Part of a sculpture is made from a solid hemisphere and a solid cylinder. The hemisphere is attached to one end of the cylinder. The cylinder has height r and radius r. Find the distance, in terms of r, of the centre of mass from the flat end of the composite solid when it is lying horizontally.

8 A cube of side 25 cm has a smaller cube of side 5 cm cut out of the centre of its front face. Assuming that the front face lies in or on the x–y plane, find its centre of mass.

Mathematics in life and work: Group discussion

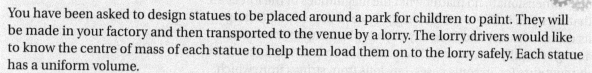

You have been asked to design statues to be placed around a park for children to paint. They will be made in your factory and then transported to the venue by a lorry. The lorry drivers would like to know the centre of mass of each statue to help them load them on to the lorry safely. Each statue has a uniform volume.

1 The first statue consists of a cylinder, with a base of radius 50 cm and height of 1.5 m, with a hemisphere of radius 50 cm attached to the top. How far from the base is its centre of mass?

2 The second statue is a cuboid, with base 40 cm × 40 cm and height 1.5 m, with a square-based pyramid that fits exactly on to the top of the cuboid and has height 0.75 m. How far from the base is its centre of mass?

3 The final statue is a cube with sides 1.5 m. It has a cuboid of size 0.3 m × 0.3 m × 1.5 m cut out of the bottom two corners. Its cross-section is shown below. Find the position of its centre of mass.

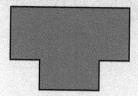

4 Which of the statues do you think will be the most stable when it is being transported? Why?

2.5 Horizontal rods

Consider a plank of wood balanced upon two logs.

For a system to be in equilibrium, not only do the forces total zero in any direction, but the moments must also balance (i.e. all the clockwise moments must balance with all the anticlockwise moments).

A standard plank will have its mass spread out evenly so its centre of mass is assumed to act at the middle of the plank. In this case, the plank is described as uniform. Alternatively, if the centre of mass is not positioned at the middle of the plank, or the mass is not evenly spread out, then it is described as non-uniform. However, it is possible for a non-uniform object, such as the dumbbells that weightlifters lift, to have its centre of mass at the middle.

KEY INFORMATION

For a system to be in equilibrium:
- the forces must total zero in any direction
- the moments must balance.

KEY INFORMATION

A uniform rod has its centre of mass at the middle of the rod.

For a non-uniform rod, the centre of mass might be at the middle, but is usually somewhere else.

According to Newton's third law, there must be an equal and opposite force to balance the weight of an object. In the example of the plank, this is provided by the reaction forces at the two logs. The points where the logs make contact with the plank are analogous to the point where the ground makes contact with a particle, where there is a normal reaction force acting perpendicular to the surface.

The plank is modelled as a rod, which is defined as rigid (and hence one-dimensional), no matter what the magnitudes of the forces are that act upon it. There will be weight forces from the rod (acting at its centre of mass) and from any objects placed upon the rod, all acting vertically downwards. There will also be reaction forces from the supports (or, in some cases, tensions from strings from which the rod is suspended) acting vertically upwards. The force diagram for the plank and logs would be drawn as shown below, with reaction forces at A and B and a weight force at the middle.

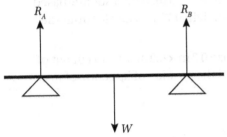

> **KEY INFORMATION**
>
> Reaction forces act vertically on the other side of the rod from the support.
>
> Weight forces act vertically downwards.

Consider a uniform rod AD of length $10\,\text{m}$, with a centre of mass at C, supported at B and D. $AB = 3\,\text{m}$. Assume that the weight of the rod is $42\,\text{N}$ and that you need to find the magnitudes of the reaction forces at B and D.

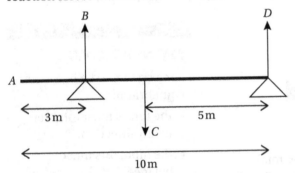

Since you are told the rod is uniform, the $42\,\text{N}$ weight is modelled as a single force acting at the middle of the rod (and since the rod is $10\,\text{m}$ long, this will be $5\,\text{m}$ from each end).

You do not know either reaction force. You could take moments about two points chosen at random (assuming these are neither B nor D) and solve two simultaneous equations. However, if you take moments about B or D, then the moment of the reaction force about that point will be zero (since the perpendicular distance is zero). Eliminating a force from the equation means that there will only be one unknown rather than two. Note that it is also easier to take moments about the end of a rod than about a point in the middle because all the forces with moments of the same sense will be pointing in the same direction (upwards or downwards).

Since there is a reaction force at D, and D is at the end of the rod, this is a sensible choice of point to take moments about.

Turning moment = force × perpendicular distance

The reaction at B is acting clockwise with a moment of $R_B \times 7$.

The weight at C is acting anticlockwise with a moment of 42×5.

Hence overall moment = $42 \times 5 - R_B \times 7 = 0$

$$210 = R_B \times 7$$

$$R_B = 30\,\text{N}$$

To find the reaction force at D, you could now take moments about any other point since you know the value of R_B. However, the simplest approach is to resolve vertically. Resolving vertically, the sum of all the upwards forces equals the sum of all the downwards forces.

$$R_B + R_D = 42$$
$$R_D = 42 - R_B$$
$$R_D = 42 - 30$$
$$R_D = 12\,\text{N}$$

Example 9

A uniform rod AB, 8 m long, is in equilibrium and rests upon supports at C and B. $AC:CB = 1:3$.

The reaction force at B is $2g\,\text{N}$ and there is a mass of 5 kg at A.

Find:

a the mass of the rod

b the reaction at C.

Solution

a Start by drawing a diagram to represent the information. Because AB is 8 m long, with C dividing it in the ratio 1:3, AC is 2 m and CB is 6 m. Since the rod is uniform, the weight force W will act at the middle of the rod. There will be a reaction force at each support including $2g\,\text{N}$ at B and a second weight force of $5g\,\text{N}$ at A.

Where there are multiple unknowns, it helps to take moments about the position at which an unknown force is acting because then the unknown force will not feature in the equation and there will be fewer variables. The unknown force in this example is at C.

45

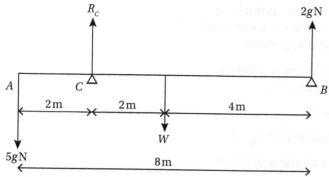

Since you do not know the reaction at C, eliminate this by taking moments about C. W is 2 m from C, the reaction at B is 6 m from C and the 5 kg mass is 2 m from C.

$$2g \times 6 + 5g \times 2 - W \times 2 = 0$$

$$2W = 22g$$

$$W = 11g\,\text{N}$$

Hence the mass of the rod is 11 kg.

b Resolving vertically:

$$R_C + R_B - 5g - W = 0$$

$$R_C + 2g - 5g - 11g = 0$$

$$R_C = 14g = 140\,\text{N}$$

The reaction force at C is 140 N.

Example 10

A horizontal non-uniform tree trunk AB, of length 36 m, rests on a support at C where $AC = 9$ m. A woman of mass 48 kg sits at A and a boy of mass 12 kg sits at B. The system is in equilibrium.

a Suppose the trunk of the tree weighs 6 kg. Find the position of its centre of mass.

b Suppose now that the mass of the trunk is unknown. What would its mass be if the question had told you the centre of mass was 12 m from A?

c What assumptions have you made in parts **a** and **b**?

Solution

a Draw a diagram, representing the tree trunk as a rod and the woman and boy as particles (with mass concentrated at a single point). Because the tree trunk is non-uniform, the centre of mass of the trunk does not have to be in the middle. For part **b**, you are told its position. For part **a**, however, just draw an arrow for the weight force at a distance of x m from A. Note that the weight force is $6g\,\text{N}$.

> If the rod is non-uniform and you have not been given the position of the centre of mass, assign a position yourself.

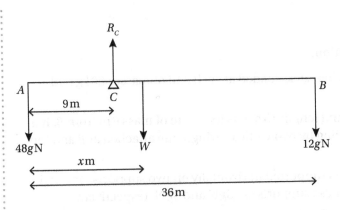

Take moments about C to eliminate the reaction force:

$$48g \times 9 - 6g \times (x - 9) - 12g \times 27 = 0$$

$$432g = 6g(x - 9) + 324g$$

$$108g = 6g(x - 9)$$

Divide by 6g:

$$x - 9 = 18$$

$$x = 27$$

The centre of mass is 27 m from A.

> Drawing a force diagram will help you to visualise the problem.

b Use the same diagram but the centre of mass is now 12 m from A and it is the weight force that is now unknown.

Again, take moments about C to eliminate the reaction force:

$$48g \times 9 - W \times 3 - 12g \times 27 = 0$$

$$108g = 3W$$

$$W = 36g$$

The mass of the trunk is 36 kg.

> When drawing the diagram for this problem, you did not know where the centre of mass of the tree was, so it was placed in an arbitrary position. You now know that $x = 27$. Although the diagram looks inaccurate, the numerical answer is correct.

c The assumptions made in this question are that the tree trunk can be modelled as a rod and the woman and boy as particles.

Stop and think   Why is there no reaction force acting on the woman or the boy?

Exercise 2.5A

1 Draw a diagram to represent each situation.

a An 8 m uniform pole weighs 30 kg. It has sacks loaded at each end of masses 20 kg and 40 kg. It rests on one support.

b A non-uniform rod AB of mass 9 kg and length 15 m has its centre of mass 6 m from B. It is suspended horizontally from the ceiling by two vertical strings, one attached at B and the other at C, where $BC = 12$ m.

c A non-uniform rod AB of length 6 m is supported horizontally on two supports, one at A and the other at B. The reactions at these supports are $5g$ N and $3g$ N, respectively.

d A 12 m uniform pole AB of mass 20 kg has supports at C and D. $AC:CD:DB = 1:3:2$.

e A playground seesaw consists of a uniform beam of length 4 m supported at its midpoint. A girl of mass 25 kg sits at one end of the seesaw. Her brother of mass 40 kg sits on the other side and the seesaw is horizontal.

f A pole vaulter uses a uniform pole of length 5 m and mass 7 kg. He holds the pole horizontally by placing one hand at one end of the pole and the other hand 80 cm away from the end of the pole.

2 A uniform log AB of mass 60 kg has loads of 30 kg and 50 kg at A and B, respectively, as shown in the diagram. The log is held by two supports at C and D, with $AC = 2$ m, $CD = 2$ m and $DB = 1$ m.

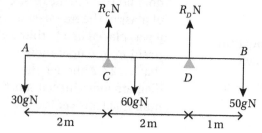

a Explain why the log is modelled as a uniform rod.

b Find the magnitudes of the reaction forces at C and D.

3 AB is a non-uniform rod of length 5 m and mass 20 kg. It rests upon a single support C where $CB = 4$ m. There is a 30 kg rock at A and a 2 kg stone at B.

Find:

a the distance of the centre of mass from A

b the reaction at C.

4 A pole vaulter uses a uniform pole of length 4 m and mass 5 kg. He holds the pole horizontally by placing one hand at one end of the pole and the other hand 80 cm away from the end of the pole. Find the vertical forces exerted by his hands.

5 A uniform beam JK of length 12 m and mass 16 kg has a sack of mass 24 kg attached at J and a sack of mass 40 kg attached at K. The beam is balanced horizontally on a support at L.

a By taking moments about L, prove that JL is 7.2 m.

b Explain why the reaction at L is 800 N.

c What assumptions have you made in answering this question?

6 Three uniform rods, each of length 25 cm but of masses 3 kg, 5 kg and 7 kg, are joined to make one rod 75 cm long, with the 5 kg rod in the middle. The rod is suspended from a vertical string attached to the rod at a point which is d cm from its centre. The rod is horizontal.

Find:

a the value of d

b the tension in the string.

7 A lumberjack has rested a tree trunk AC of mass 64 kg and length 6 metres on two supports, one at A and one at B. When a 16 kg boulder is placed on the tree trunk at C, the reaction force at B is three times the reaction force at A. The tree is assumed to be uniform.

a What is the reaction force at B?

b Find the position of B.

The 16 kg boulder is removed and replaced by a boulder of mass M kg. The reaction force at B is now five times the reaction force at A.

c Show that M is 32.

d What assumption have you made about the boulders and why did you need to make this assumption?

8 A non-uniform rod CD, 10 m long, is supported at A and B, where $AC = 3$ m and $BD = 2$ m. When a 10 kg object is placed at D, the ratio of the reactions at A and B is 1:4, but when a 5 kg object is placed at C instead, the ratio of the reactions at A and B is 4:3.

a Find the weight of the rod.

b Find the distance of the centre of mass of the rod from C.

Tilting

Consider a plank of wood balanced on two logs. If you were to stand on one end of the plank, unless you were very light or the plank was very heavy, the plank would be likely to tilt.

The diagram shows a rod resting on two supports, P and Q.

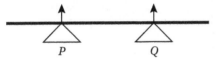

If the rod rotates about P, then the rod will no longer be in contact with Q. At the instant that the rod starts to rotate, the rod is described as being on the point of **tilting** about P, and the reaction force at Q is zero.

> **KEY INFORMATION**
>
> If a rod is supported at P and Q and the rod is on the point of tilting about P, then $R_Q = 0$.

Example 11

A uniform rod AB, of length $18\,\text{m}$, rests on supports at C and D, where $AC:CD:DB = 1:3:2$.

It is on the point of tilting about D. There is a mass of $3\,\text{kg}$ at A and a mass of $8\,\text{kg}$ at B.

Find the weight of the rod and the reaction at D.

Solution

Dividing $18\,\text{m}$ in the ratio $1:3:2$, $AC = 3\,\text{m}$, $CD = 9\,\text{m}$ and $DB = 6\,\text{m}$. The rod is uniform, so the weight of the rod is $9\,\text{m}$ from A. Because the rod is on the point of tilting about D, the reaction force at C is zero.

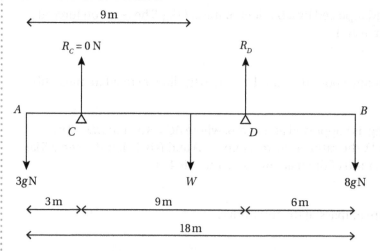

Even though the reaction at C is zero, it is still advisable to include the force in the diagram.

Take moments about D to eliminate the reaction force at D. D is $3\,\text{m}$ from the centre of mass.

$$3g \times 12 + W \times 3 - 8g \times 6 = 0$$

$$36g + 3W = 48g$$

$$3W = 12g$$

$$W = 4g = 40\,\text{N}$$

The weight of the rod is $40\,\text{N}$.

Resolving vertically:

$$R_D + R_C - 3g - W - 8g = 0$$

$$R_D + 0 - 3g - 4g - 8g = 0$$

$$R_D = 15g = 150\,\text{N}$$

The reaction force at D is $150\,\text{N}$.

Exercise 2.5B

 1 A uniform beam *AB*, of mass 6 kg, is suspended from two vertical strings attached at points *J* and *K*, where *AJ* is *d* m. The beam is 4*d* m in length and horizontal. A particle of mass 9 kg is attached to the beam and, as a result, the beam is now on the point of tilting about *J*. How far is the particle from *A* in terms of *d*?

 2 A non-uniform rod *AB* has a mass of 10 kg and length of 2.1 m. It rests on supports at *C* and *D*, where *AC*, *CD* and *DB* are equal distances. The rod is horizontal and in equilibrium. When a particle of mass 4 kg is attached to the rod at *B*, the rod is on the point of tilting about *D*. Show that the centre of mass is 1.12 m from *A*.

 3 A uniform rod *CF* has a mass of 70 kg. It rests on supports *D* and *E*, which are 3 m and 7 m from *C*, respectively, as shown in the diagram. The rod is on the point of tilting about *D*. There is a 50 kg particle at *C* and an *m* kg particle at *F*.

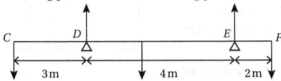

   a State the magnitude of the reaction force at *E*.

   b Find the value of *m*.

 4 A telegraph pole *AB* has length 12 m and mass 20 kg. It is resting in equilibrium in a horizontal position, on supports at points *P* and *Q*, where *AP* = 4 m and *QB* = 3 m. The telegraph pole is modelled as a uniform rod.

   a Find the reaction forces at *P* and *Q*.

   A crate of mass *M* kg is placed at point *X* where *AX* = 10 m. The rod is now on the point of tilting about *Q*.

   b Find the value of *M*.

 5 The diagram shows a uniform plank *XY* of length 10 m and weight *W* N, balanced on supports at *U* and *V*. There is a 70 kg load 1 m from *X* and a 5 kg load at *Y*. *XU* = 2 m and *XV* = 7 m.

   a Given that the plank is on the point of tilting about *U*, find the value of *W*.

   b Another load is added at *Y* so that the rod is now on the point of tilting about *V*. Show that the mass of the load that has been added can be written in the form $\frac{a}{3}$ kg, where *a* is an integer.

 6 A non-uniform 8 m rod *AB* resting on supports at *E* and *F* has a weight of *Mg*. *AE* = 2 m and *AF* = 5 m. In this situation, the reaction at *F* is 18*g* N but when a 72 kg block is placed 1.5 m from *B*, the rod is on the point of tilting about *F*. Find the value of *m* and the distance of the rod's centre of mass from *A*.

2.6 Equilibrium of rigid bodies

In the previous section, all the forces acted vertically on a horizontal rod. In this section, the forces can act in any direction.

Ladders

When considering a ladder leaning against a wall, all the forces are horizontal or vertical. The ladder rests against a wall and experiences a horizontal reaction from the wall. If the wall is rough, there will also be a friction force vertically upwards. The ladder rests on the ground and again there will be a reaction force, but this time it will be vertical. There must be a force keeping the ladder from slipping – if the floor is rough this will be a horizontal friction force acting towards the wall. However, there could be a taut string keeping it in place. The ladder will also have a weight.

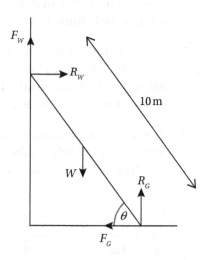

This diagram shows the weight and both of the friction and reaction forces for a 10 m length ladder inclined at an angle of θ to the horizontal.

Problems on ladders are solved by a combination of resolving and taking moments.

For the diagram above:

Resolving horizontally: $R_W - F_G = 0$

Resolving vertically: $R_G + F_W - W = 0$

Taking moments about the point of contact of the ladder with the ground:

$W \times 5 \cos \theta - F_W \times 10 \cos \theta - R_W \times 10 \sin \theta = 0$

The three diagrams below show the perpendicular distances for the friction force at the wall, the reaction force at the wall and the weight.

> **KEY INFORMATION**
>
> Solve ladder problems by using simultaneous equations. Generate equations by resolving and using moments and $F \leq \mu R$.

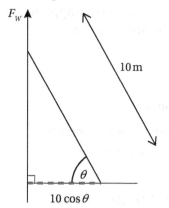

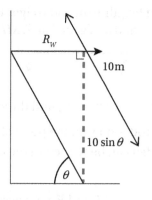

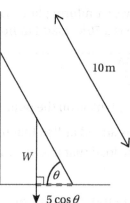

Example 12

A uniform ladder of mass 30 kg rests against a smooth wall at an angle of arctan 3 to the horizontal. The ground is rough. Given that the ladder is in limiting equilibrium, find the coefficient of friction between the ladder and the ground.

> If the ladder is uniform then its weight will act downwards from the centre of the ladder.

Solution

Draw a diagram. Let the point of contact with the wall be W and the point of contact with the ground be G. The ground is rough so has both a reaction and a friction force whereas the wall is smooth so only has a reaction force. The ladder is uniform so its weight acts at the middle. The ladder is inclined at an angle of arctan 3 to the horizontal. Let the angle be α, where $\tan \alpha = 3$.

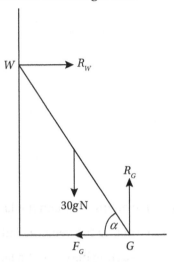

> **KEY INFORMATION**
>
> Use $F = \mu R$ for limiting equilibrium.

> If a length is not given, then it may not be needed. By choosing the length of the ladder to be equal to $2a$, it allows you to use a as the distance between the weight and each end, instead of the fraction $\frac{a}{2}$.

The length of the ladder is not given. Let the length of the ladder equal $2a$ m, so the weight acts at a m from each end.

Since there are two forces acting at G and only one at W, take moments about G.

The perpendicular distance between G and the line of action of R_W is $2a \sin \alpha$ m.

The perpendicular distance between G and the line of action of $30g$ N is $a \cos \alpha$ m.

> In this example we have used force × perpendicular distance, but you would get the same outcome by using distance × perpendicular force.

Hence $30g \times a \cos \alpha - R_W \times 2a \sin \alpha = 0$

Dividing by $a \cos \alpha$: $30g - R_W \times 2 \tan \alpha = 0$

Since $\tan \alpha = 3$: $30g = R_W \times 2 \times 3$

$$R_W = 5g \, \text{N}$$

Resolving horizontally and vertically:

$R_W - F_G = 0$

$R_G - 30g = 0$

The question asks for the coefficient of friction between the ladder and the ground. Since the ladder is in limiting equilibrium, you have $F_G = \mu R_G$, from which $\mu = \dfrac{F_G}{R_G}$.

$\mu = \dfrac{F_G}{R_G} = \dfrac{R_W}{30g} = \dfrac{5g}{30g} = \dfrac{1}{6}$

Exercise 2.6A

 1 A ladder of mass M kg is modelled as a uniform rod. The wall is smooth but the floor is rough. The coefficient of friction acting between the ladder and the floor is 0.4. The ladder is inclined at an angle of α to the vertical and is on the verge of slipping.

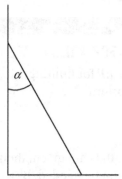

Find:

a the magnitude of the reaction force acting between the ladder and the floor in terms of M

b the magnitude of the friction force acting between the ladder and the floor in terms of M

c the magnitude of the reaction force acting between the ladder and the wall in terms of M

d the angle α.

 2 A non-uniform ladder AB of mass 50 kg and length 8 m stands on rough ground, resting against a smooth wall with the point of contact at A.

The angle between the ground and the ladder is 70°. The coefficient of friction acting between the ladder and the ground is $\frac{1}{4}$ and the ladder is on the verge of slipping.

Find the distance of the centre of mass of the ladder from B.

 3 A uniform ladder GW is 6 m long, has a mass of 36 kg and makes an angle of 30° with the vertical. G is the point of contact of the ladder with the ground and W is the point of contact of the ladder with the wall. A 45 kg boy stands on the ladder 2 m from G. The wall is smooth. Show that the coefficient of friction acting between the ladder and the ground is given by $\frac{11\sqrt{3}}{81}$.

 4 A uniform ladder CD of mass 54 kg rests against a rough wall, where C is on the ground.

The length of the ladder is $2a$ metres.

The ladder is at an angle of α to the horizontal, where $\sin\alpha = 0.8$.

The coefficient of friction acting between the ladder and the wall is $\frac{1}{6}$.

Show that the coefficient of friction acting between the ladder and the ground when the ladder is on the verge of slipping can be written as $\frac{k}{17}$, where k is a constant to be found.

 5 A ladder AB of mass 60 kg and length 3.5 m leans against a smooth wall. The ladder is inclined at an angle α to the horizontal, where $\tan\alpha = \frac{20}{13}$. The foot of the ladder, B, is on rough horizontal ground.

The coefficient of friction acting between the ladder and the ground is 0.4. The ladder is modelled as a uniform rod. A man of mass 70 kg climbs the ladder. Find the maximum distance he can climb safely.

6 A uniform ladder *CD* of mass 40 kg rests against a rough wall, where *C* is on the ground.

The length of the ladder is 2*a* metres.

The ladder is at an angle of α to the horizontal where $\sin \alpha = 0.8$.

The coefficient of friction acting between the ladder and the wall, μ, is the same as the coefficient of friction acting between the ladder and the ground.

Find the value of μ.

Stop and think   What is the minimum angle a ladder can make with the horizontal? Why is this?

Other bodies

You will now consider other rigid bodies in which one or more forces will act at an angle (i.e. not all the forces will be parallel or perpendicular). In this case, it will be necessary to resolve a force into its components.

Often there will still be a reaction force and a friction force acting at the point of contact and related by the relationship $F \leqslant \mu R$ (or $F = \mu R$ for limiting equilibrium).

Alternatively, where a rod is connected to a surface by a hinge, there will be a reaction force acting between the rod and the hinge. Since the hinge is a point rather than a surface, the angle cannot be considered to be perpendicular. In this case, the reaction force needs to be resolved into *X* and *Y* components. The magnitude of the reaction force can then be deduced by using Pythagoras' theorem and the angle by trigonometry.

KEY INFORMATION

Use *X* and *Y* for the components of the reaction at a hinge, since the angle is unknown.

Example 13

The diagram shows a uniform pole *AB*, of mass 8 kg and length 2.5 m, freely hinged at *A* to a vertical wall. The pole is held horizontally in position by a light inextensible rope *BC*, with *C* vertically above *A*. The string is inclined at angle α to the horizontal, such that $\tan \alpha = \frac{2}{3}$.

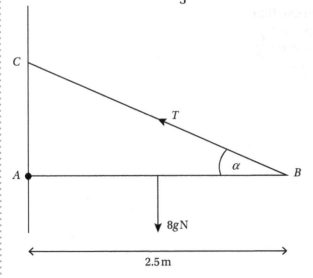

a Show that the tension in the rope is given by $T = 2\sqrt{13}\,g$ N.

A load of mass 5 kg is now attached to the pole at D, where $BD = 0.5$ m.

b Find the magnitude and direction of the force exerted by the hinge on the pole at A.

Solution

a Start by completing the diagram, with the reaction force acting between the pole and the hinge. Since you do not know the angle at which the reaction acts, represent it as two perpendicular components, X for the horizontal and Y for the vertical.

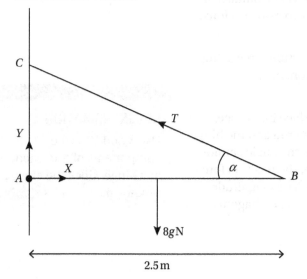

Take moments about A to eliminate both X and Y from the equation. This requires you to resolve T into components. The perpendicular component is $T \sin \alpha$.

$$8g \times 1.25 = T \sin \alpha \times 2.5$$

Given that $\tan \alpha = \frac{2}{3}$ (and that α is acute), you can find the value of $\sin \alpha$ using a right-angled triangle and trigonometry. For any right-angled triangle, $\tan \alpha = \frac{O}{A}$. If the opposite is 2 and the adjacent is 3, then the hypotenuse is $\sqrt{2^2 + 3^2} = \sqrt{13}$.

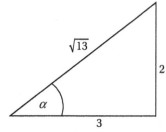

Hence $\sin \alpha = \frac{O}{H} = \frac{2}{\sqrt{13}}$.

$$8g \times 1.25 = T \times \frac{2}{\sqrt{13}} \times 2.5$$
$$10g = T \times \frac{5}{\sqrt{13}}$$
$$T = 10g \times \frac{\sqrt{13}}{5}$$
$$T = 2\sqrt{13}g \, \text{N}$$

b A load of mass 5 kg has been added at D, so add this to the diagram.

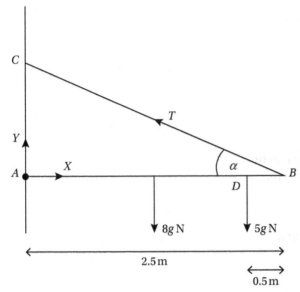

Take moments about A to find the new tension:

$$T \sin \alpha \times 2.5 - 8g \times 1.25 - 5g \times 2 = 0$$
$$T \times \frac{5}{\sqrt{13}} = 20g$$
$$T = 4\sqrt{13}g \, \text{N}$$

Resolve horizontally to find X:

$$X - T \cos \alpha = 0$$

You can use the same right-angled triangle as before to find the value of $\cos \alpha$.

$$\cos \alpha = \frac{A}{H} = \frac{3}{\sqrt{13}}$$
$$X = 4\sqrt{13}g \times \frac{3}{\sqrt{13}} = 12g \, \text{N}$$

Resolve vertically to find Y:

$$Y + T \sin \alpha - 13g = 0$$
$$Y + 4\sqrt{13}g \times \frac{2}{\sqrt{13}} = 13g$$
$$Y + 8g = 13g$$
$$Y = 5g \, \text{N}$$

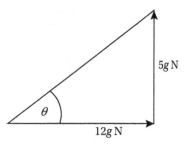

5g N

12g N

θ

Using the triangle law, you can add the components X and Y. Since X and Y are perpendicular, the magnitude of the reaction force R can be calculated using Pythagoras' theorem.

$R = \sqrt{(5g)^2 + (12g)^2}$

$R = \sqrt{25g^2 + 144g^2}$

$R = \sqrt{169g^2}$

$R = 13g = 130\,\text{N}$

Similarly, the direction of the reaction force can be found using the same triangle and trigonometry.

$\tan\theta = \dfrac{5g}{12g} = \dfrac{5}{12}$

$\theta = \tan^{-1}\left(\dfrac{5}{12}\right) = 22.6°$ above the horizontal.

Example 14

The diagram shows a non-uniform beam AB, of mass 5 kg and length 4 m, resting against a peg at C, where AC = 3 m. The end A of the beam lies on a rough horizontal floor. The angle between the beam and the floor is 30°. The reaction force exerted on the beam by the peg is $14\sqrt{3}\,\text{N}$. The beam is in limiting equilibrium. The centre of mass of the beam is d m from A.

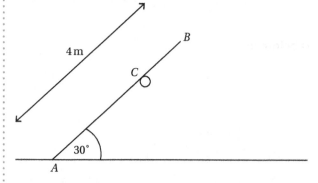

B

4 m

C

30°

A

a Find the value of d.

b Show that the coefficient of friction acting between the beam and the floor at A is given by $k\sqrt{3}$, where k is a constant to be found.

Solution

a Start by adding extra information to the diagram. The peg is 1 m from B (3 m from A). The reaction force at the peg is $14\sqrt{3}$ N and acts perpendicular to the beam. There will be a reaction force, R_A, and a friction force, F, at A.

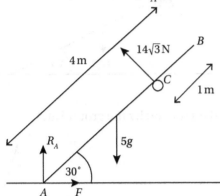

Take moments about A to eliminate R_A and F:

$$14\sqrt{3} \times 3 - 5g \times d\cos 30° = 0$$

$$d = \frac{14\sqrt{3} \times 3}{5g\cos 30°} = 1.68\,\text{m}$$

b Resolve horizontally to find the friction force at A:

$$F - 14\sqrt{3} \times \sin 30° = 0$$

$$F = 14\sqrt{3} \times \frac{1}{2}$$

$$F = 7\sqrt{3}\,\text{N}$$

Resolve vertically to find the reaction force at A:

$$R_A + 14\sqrt{3} \times \cos 30° - 5g = 0$$

$$R_A + 14\sqrt{3} \times \frac{\sqrt{3}}{2} = 5g$$

$$R_A + 21 = 50$$

$$R_A = 29\,\text{N}$$

Since the beam is in limiting equilibrium, $F = \mu R$ and $\mu = \dfrac{F}{R}$.

So $\mu = \dfrac{7\sqrt{3}}{29}$.

Hence $k = \dfrac{7}{29}$.

Exercise 2.6B

(PS) 1 A uniform rod AB, of mass 12 kg and length 1.8 m, rests against a peg at C, where $AC:CB = 2:1$. The rod is inclined at an angle of 25° to the horizontal with A at rest on rough horizontal ground.

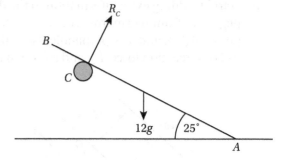

a Find the reaction force exerted on the rod by the peg.

b Find the frictional force at A.

c Find the reaction force at A.

d Find the magnitude and direction of the force exerted on the rod by the ground at A.

(MM) 2 A thin metal pole AB is modelled as a uniform rod with a mass of 8 kg and a length of 1.6 m. The pole is freely hinged at A to a vertical wall and held horizontally by a light inextensible cable at B. The cable makes an angle of 60° with the horizontal, as shown in the diagram.

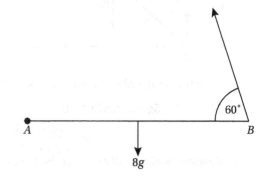

a Find the tension in the cable.

b Find the magnitude and direction of the force exerted by the hinge on the rod.

(PS) (C) 3 A non-uniform pole AB, of mass M kg and length $12a$ m, is held horizontally in position, with A at rest on a rough vertical wall, by a light inextensible string BC of length $13a$ m. C is vertically above A. The coefficient of friction between the wall and the pole is given by μ. The centre of mass of the pole is $4a$ m from A.

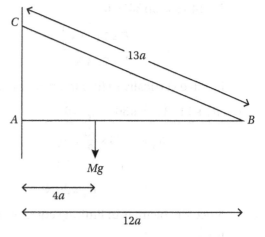

a Show that the tension in the string is $\dfrac{13}{15} Mg$ N.

b Show that $\mu \geqslant \dfrac{5}{6}$.

(C) 4 $ABCD$ is a uniform rectangular sheet of metal of mass 6 kg. The length of AB is $2a$ m and of BC is $10a$ m. The lamina is hinged at A and is able to swing freely. The sheet is held in place with AD horizontal by a force of $2F$ N acting along BC and a force of F N acting along CD. Show that the reaction force exerted on the sheet by the hinge acts at an angle of arctan 0.9 to the horizontal.

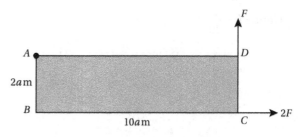

5 A plank AB of mass 7.5 kg is supported at rest by a peg at C. The reaction force exerted on the plank by the peg is 49 N. The end A rests on rough horizontal ground. The plank is in limiting equilibrium and the coefficient of friction between the ground and the plank is μ. AB is 3 m in length. The peg is 2.5 m from A and the centre of mass of the plank is 2 m from A. The plank is inclined at angle α to the horizontal.

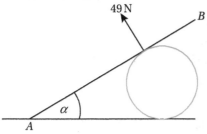

a Find the size of the angle α, correct to 1 decimal place.

b Find the value of μ.

6 A plastic bar AB of mass 0.6 kg is hinged at point A on a vertical wall. The bar is held in a horizontal position by a solid strut CD, where C is vertically below A. The angle between the wall and the strut is 75°. The length of AD is $\frac{3}{4}$ of the length of AB. The bar is modelled as a uniform rod.

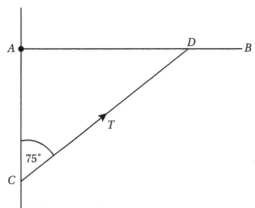

a Show that the thrust in the strut is 15.5 N, correct to 3 significant figures.

b Find the magnitude of the force exerted on the bar by the hinge.

7 A rod AB, of mass 2 kg and length 3 m, is held horizontally, as shown in the diagram, by two light inextensible ropes. The rope at A is inclined at 20° to the horizontal and the rope at B is inclined at 50° to the horizontal. The centre of mass of the rod is x m from A.

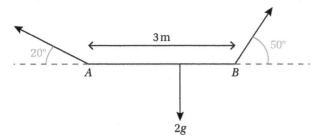

a Find the magnitude of the tension of each rope.

b Find the value of x.

2.7 Determining whether an object will topple or slide

In this section, you will look at what happens when an object on a plane surface is not in equilibrium. If it is not in equilibrium, it will either slide or topple. You need to be able to work out which of these will happen first.

To do this, you need to have a clear diagram showing all of the forces parallel and perpendicular to the surface. Make sure you include the weight (W) of the object, the normal reaction force (R), the friction force (F) and any applied force.

If an object is on the point of **sliding**, then $F = \mu R$.

If an object is on the point of **toppling**, then its centre of mass is directly above the lowest point of contact between the object and the plane.

Example 15

A force, P, is applied to a cuboid of mass 10 kg until it moves, as shown below. The coefficient of friction between the cuboid and the plane is 0.5. If the force P is gradually increased, will the cuboid slide or topple first?

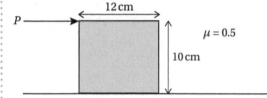

Solution

The first step is to add all of the forces to the diagram.

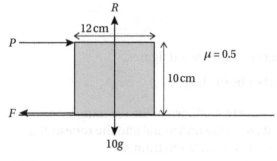

> If P was applied at an angle, you would need to resolve the forces so that they are parallel and perpendicular to the plane surface.

Sliding

Resolving horizontally: $P = F$

Resolving vertically: $R = 10g$

At the point of sliding, $F = \mu R$

$$= 0.5 \times 10g$$
$$= 50\,\text{N}$$

So the cuboid will be on the point of sliding when the applied force is 50 N.

Toppling

If the cuboid were to topple, point A would be the pivot point. In this case, the reaction force would pass through A.

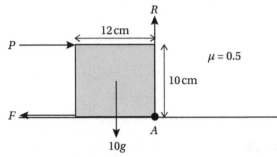

Taking moments about A:

Moments $(A) = 6 \times 10g - 10 \times P$

Since the cuboid is on the point of toppling, the total of the moments of the system must be 0.

So $6 \times 10g - 10 \times P = 0$

$$P = 60\,\text{N}$$

So the cuboid will be on the point of toppling when the applied force is 60 N.

Compare the two answers – the smaller of the two values is 50 N. This means that the cuboid will slide first.

> Forces R and F pass through A so their moments are 0.

Example 16

A uniform cube of side 8 cm and mass 16 kg is placed on a slope that is inclined at an angle α above the horizontal. The coefficient of friction between the slope and the cube is 0.45. If α is gradually increased, will the cube slide down the slope or topple first?

Solution

First draw a diagram showing all of the forces parallel or perpendicular to the plane.

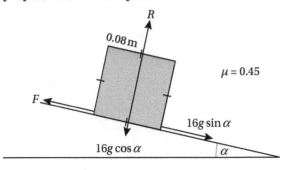

Sliding

Resolving parallel to the slope: $F = 16g \sin \alpha$

Resolving perpendicular to the slope: $R = 16g \cos \alpha$

At the point of sliding, $F = \mu R$

$$16g \sin \alpha = 0.45 \times 16g \cos \alpha$$

$$\tan \alpha = \frac{0.45 \times 16g}{16g}$$

$$= 0.45$$

$$\alpha = \arctan(0.45)$$

$$= 24.2° \text{ (to 1 d.p.)}$$

So the cube will be on the point of sliding when the angle reaches 24.2°.

Toppling

When the cube is on the point of toppling, the weight will pass through the lowest vertex of the cube. This can be seen in the diagram below.

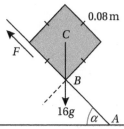

Looking only at the cube, you can form a right-angled triangle with angle α at the base and BC as the hypotenuse, when C is the centre of the square. The opposite and adjacent are each 0.04 m long.

So $\tan \alpha = \dfrac{0.04}{0.04}$

$$\alpha = \arctan(1)$$

$$= 45°$$

So the cube will be on the point of toppling when the angle reaches 45°.

Compare the two answers – the smaller of the two angles is 24.2°. This means that the cube will slide first.

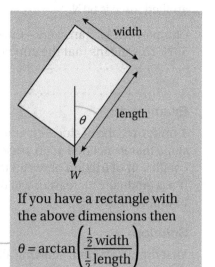

If you have a rectangle with the above dimensions then

$$\theta = \arctan\left(\frac{\frac{1}{2}\text{ width}}{\frac{1}{2}\text{ length}}\right)$$

Exercise 2.7A

 1 **a** A force, P, is applied to a cube of mass 12 kg as shown. The coefficient of friction between the cube and the plane surface is 0.32. If the cube is on the point of sliding, what force is being applied?

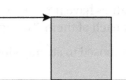

b The force is changed so that it is now applied at the same point but from an angle of 15° above the horizontal. What is the minimum force that needs to be applied for the cube to slide?

 2 A uniform cuboid of size 20 cm by 10 cm by 10 cm and mass 2 kg is standing on a smooth surface as, shown. A force, P, is applied to the top of the cuboid.

a What force is required to topple the cuboid if $\mu = 0.3$?

b What force is required to slide the cuboid if $\mu = 0.3$?

c Is the cuboid going to slide or topple first?

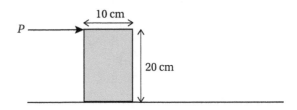

 3 A solid wooden cuboid, measuring 50 cm by 40 cm by 30 cm and of mass 2.5 kg, is lying on a table. The coefficient of friction between the cuboid and the table is 0.2.

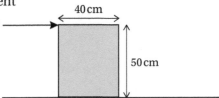

a A force, P, is applied to the cuboid, as shown, and is gradually increased from 0 N. Will the cuboid slide or topple first? You must show all of your working.

b The force is now applied from the top right edge of the block at an angle of 10° above the horizontal. Will the cuboid slide or topple first?

 4 A cuboid, measuring 10 cm by 5 cm by 5 cm and of mass 4 kg, is on a plane, smallest face down, inclined at an angle α above the horizontal. If $\mu = 0.30$, find:

a the angle α when the cuboid is on the point of sliding

b the angle α when the cuboid is on the point of toppling

c whether the cuboid will slide or topple first.

 5 Jiang is carrying out an experiment to calculate the coefficient of friction between two surfaces. He made a cuboid from one material and placed it, square side down, on an inclined surface made from the other material. The cuboid measures 24 cm by 10 cm by 10 cm and has mass 12 kg. He finds that the cuboid begins to slide when $\alpha = 20°$.

a Find the coefficient of friction between the two surfaces.

b At what value of α would the cuboid topple?

Mathematics in life and work: Group discussion

The three statues have now been delivered to the park and you have been asked to help decide where to put each of them. The park has three possible locations for the statues.

Location 1: Exposed to strong winds – the statue here needs to be the least likely to topple in a strong wind.

Location 2: On a ramp inclined at 10° above the horizontal.

Location 3: On a flat surface, sheltered from the strongest winds.

Assuming that each model weighs the same and the coefficient of friction between the statues and the ground is 0.5, use calculations to decide which statue would be best in which location.

SUMMARY OF KEY POINTS

- Turning moment = force × perpendicular distance.

- The units for turning moment are N m.

- The sense of a turning moment defines whether it is clockwise or anticlockwise – anticlockwise is positive and clockwise is negative.

- For a force F acting at a distance d m at angle of θ, the turning moment is given by $Fd \sin \theta$.

- When you try to balance an object on a point, there will only be only one place where it will balance. This point is vertically below the object's centre of mass.

- The position of the centre of mass, $(\bar{x}, \bar{y})$, of a system of particles in a plane is calculated by using $M\bar{x} = \sum_{i=1}^{i=n} m_i x_i$ and $M\bar{y} = \sum_{i=1}^{i=n} m_i y_i$.

- If a uniform lamina has an axis of symmetry then the centre of mass will lie on this axis. If it has more than one axis of symmetry, the centre of mass will lie where these axes intersect.

- To find the centre of mass of a uniform lamina made of compound shapes, you replace each individual shape by a particle of the same mass at its centre of mass and then find the overall centre of mass for the lamina, using these individual points.

- The centre of mass of a triangular lamina is $\frac{2}{3}$ of the distance along the median from any vertex.

- The centre of mass of a circular sector of radius r and angle 2α is $\frac{2r \sin \alpha}{3\alpha}$ from the centre of the circle, where α is measured in radians.

- The centre of mass of a solid hemisphere with radius r is $\frac{3}{8}r$ from its centre on the line connecting the centre of the base and the highest part of the hemisphere.

- The centre of mass of a solid cone or right pyramid with height h is $\frac{3}{4}h$ below its vertex.

- If the overall moment about a point is zero, then the clockwise moments will be equal and opposite to the anticlockwise moments. For a system to be in equilibrium, not only do the forces sum to zero in any direction, but the moments must also balance (i.e. the total of the clockwise moments must balance the total of the anticlockwise moments).

- To solve problems involving uniform rods, you assume that the weight force of a rod acts through its centre of mass.

- Ladder problems can be solved using simultaneous equations generated by resolving, finding moments and using $F \leqslant \mu R$.

- If an object is on the point of sliding, then $F = \mu R$.

- If an object is on the point of toppling, then its centre of mass is directly above the lowest point of contact between the object and the plane.

- To decide whether an object will slide or topple, you need to consider both cases and work out which one will happen first.

EXAM-STYLE QUESTIONS

1

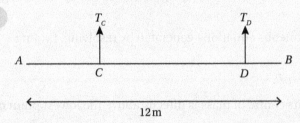

The diagram shows a non-uniform rod AB, of length 10 m and mass 90 kg. The centre of mass of AB is x m from A. There are loads of 60 kg and 40 kg at A and B. C is 1 m from A and D is 3 m from B. The reaction force at D is $95g$ N. AB rests in equilibrium.

a Find the magnitude of the reaction force at C.

b Find the value of x.

2 A non-uniform plank AB has length $5d$ m and mass $8m$ kg. It initially rests in a horizontal position on supports at the points U and V, where $AU = 2d$ m and $AV = 4d$ m. A parcel of mass $6m$ kg is placed on the plank at B. The plank is on the point of tilting about V. Find the distance of the centre of mass of AB from A.

3 A uniform rod WZ of length 14 m and mass 5 kg rests horizontally in equilibrium on two supports at X and Y, where $WX:XZ = 1:6$. A particle of mass 7 kg is attached at W and a particle of mass 9 kg is attached at Z. The reaction force at X has half the magnitude of the reaction force at Y. Find the distance WY.

4 A uniform solid is in the shape of the frustum of a cone. The shape has circular ends of radius 5 cm and 10 cm, with the larger circle forming the base. The solid shape has a height of 8 cm.

a Find the vertical distance between the base and the centre of mass of the solid.

A solid hemisphere of radius 5 cm is attached to the top of the frustum to create a new compound solid. It is given that the density of the frustum and the hemisphere are the same.

b Find the distance between the centre of mass of the frustum and the centre of mass of the new compound solid.

5

The diagram shows a uniform rod AB of length 12 m and weight W newtons. The rod is held in equilibrium in a horizontal position by two vertical ropes attached at C and D, where AC is 4 m and DB is 1 m. A particle of weight 28 N is suspended from the rod at A.

a Show that the tension in the rope at C is given by $\left(\frac{5}{7}W + 44\right)$ N.

The tension in the rope at C is 13 times the tension in the rope at D.

b Find the value of W.

6

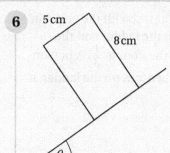

5 cm

8 cm

θ

The diagram shows a uniform cuboid positioned on a rough inclined plane. The rectangular cross-section of the cuboid that has length 8 cm and width 5 cm is in the line of greatest incline. Given that the cuboid topples before it slides, find the value of θ when the cuboid is on the point of toppling.

MM **7**

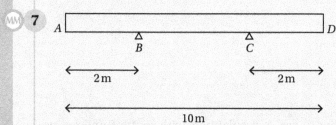

A $\quad$ D

B $\quad$ C

2 m $\quad$ 2 m

10 m

The diagram shows a non-uniform rod AD with length 10 m and weight W newtons. The rod rests horizontally on supports at B and C which are each 2 m from an end of the rod. The centre of mass of the rod is x m from A. When a 1800 N force is applied vertically upwards at A, the rod is on the point of tilting about C. When a 1575 N force is applied vertically upwards at D instead, the rod is on the point of tilting about B.

a Find the value of W.

b Find the value of x.

8

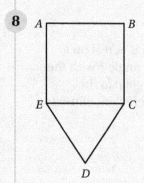

A $\quad$ B

E $\quad$ C

D

A metal frame $ABCDE$ is made from 6 uniform rods. Each rod has length 2 m.

a Find the distance of the centre of mass of the frame from D.

The frame is freely suspended from A and hangs in equilibrium.

b Find the angle that AB makes with the vertical.

9 A ladder is modelled as a uniform rod AB of mass 50 kg. The ladder is at rest with the end A in contact with a smooth vertical wall. $AB = 3.5$ m and the angle between the ladder and the ground is $\arctan 2$. The coefficient of friction between the ladder and the floor is $\frac{1}{3}$. A person of mass 70 kg climbs up the ladder from B. When the person has climbed x m up the ladder, it is on the point of sliding. Find the value of x.

10

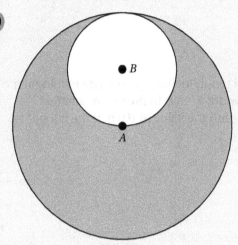

A uniform circular disc has centre A and radius $2r$ cm. A circular hole with centre B and radius r cm is made in the disc. The circumference of the circular hole passes through A and passes through the highest point on circular disc. Find the distance of the centre of mass of the object from A in terms of r.

11

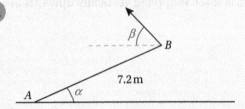

The diagram shows a uniform rod AB with length 7.2 m and mass 2.5 kg. A is at rest on a rough horizontal platform with B attached to a cable. The cable makes an angle β with the horizontal and the tension in the cable is T N. The rod is inclined at an angle α to the horizontal where $\sin \alpha = 0.6$. The rod is in limiting equilibrium and the coefficient of friction between the ground and the rod is $\frac{1}{2}$.

a Find the magnitude of the vertical reaction force on the rod at A.

b Given that $\sin \beta = 0.28$, show that $T = 17.9$ N.

c State two modelling assumptions you have made.

12 A solid uniform cylinder has radius 2 m, height 8 m and a mass of 5 kg. The cylinder is placed on a rough inclined plane, such that its circular base is flat against the plane. The incline of the plane begins small and is gradually increased. The coefficient of friction between the plane and the cylinder is 0.6. Find whether the cylinder will slide or topple down the plane first and the angle at which this occurs.

13

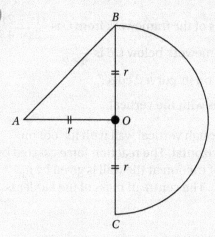

The diagram shows a uniform lamina *OABC* that is made from a triangle and a semicircle.

a Show that the horizontal distance of the centre of mass from *A* is $\dfrac{r(2+\pi)}{1+\pi}$.

b Show that the vertical distance of the centre of mass from *A* is $\dfrac{r}{3(1+\pi)}$.

The lamina is freely suspended from *A* and hangs in equilibrium.

c In the case where $r = 1 + \pi$ m, find the angle that *AO* makes with the vertical.

PS 14

The diagram shows a non-uniform rod *AB* of length 6m and mass *m* kg. The rod rests horizontally on two supports at *C* and *D*. Particles are attached to the rod at *A* and *B*. When the mass of the particle at *A* is 100 kg and the mass of the particle at *B* is 20 kg, the rod is on the point of tilting about *C*. When the mass of the particle at *A* is 20 kg and the mass of the particle at *B* is 250 kg, the rod is on the point of tilting about *D*.

a Find the value of *m*.

b Find the distance of the centre of mass of *AB* from *A*.

PS 15

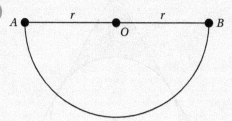

A uniform metal framework is made up of a rod *AB* and a semicircular arc *AB*, with centre *O*. A particle with the same mass as the rod is positioned at *A*. Another particle with the same mass as the semicircular arc is positioned at *B*.

a Show that the horizontal distance of the centre of mass of the framework from O is $\dfrac{r(\pi-2)}{2\pi+4}$.

b Show that the distance of the centre of mass of the framework below OB is $\dfrac{r}{\pi+2}$.

The framework is positioned so that it can balance freely on its curved edge.

c In the case where $r = 5$ m, find the angle that OB makes with the vertical.

16 A ladder of mass M kg and length $8a$ m stands against a rough vertical wall with its foot on rough ground. The ladder is inclined at angle θ to the horizontal. The reaction force exerted by the ground on the ladder is given by R_G. The coefficient of friction at the wall is given by μ_W and the coefficient of friction at the ground is given by μ_G. The centre of mass of the ladder is $6a$ m from the foot of the ladder.

a Find R_G in terms of M, μ_W and μ_G.

b Find R_G in terms of M, μ_G and θ.

c Show that $\mu_W = \dfrac{3 - 4\mu_G \tan\theta}{\mu_G}$.

17

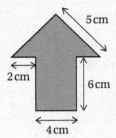

The diagram shows a prism of uniform mass in the shape of a symmetrical arrow. The thickness of the prism is 4 cm and its mass is 8 kg.

a Find the distance of the centre of mass from the centre of the apex of the triangle.

The incline of the rough ground that the prism is resting on is gradually increased from 0°. The coefficient of friction between the slope and the prism is 0.25.

b Find whether the prism will slide or topple down the slope first and the angle at which this occurs.

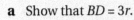
18 The diagram shows a uniform lamina made from an equilateral triangle ABC enclosing a circular hole with centre O. The circular hole has radius r m and the circumference of the circle meets the triangle at the midpoints of AB, BC and AC. D is the midpoint of AC. The centre of mass of the uniform lamina is O.

a Show that $BD = 3r$.

E is the midpoint of AD. The lamina is freely suspended from E and hangs in equilibrium.

b Find the angle between ED and the vertical.

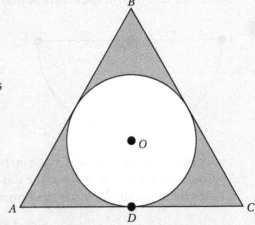

19 A uniform solid of mass 5 kg is made up of a hemisphere and a cylinder joined by a common circle of radius 4 cm. The solid rests on its flat base and has a height of 14 cm.

a Find the distance of the centre of mass from the centre of the flat base of the solid.

The incline of the rough ground that the solid is resting on is gradually increased from 0°. The coefficient of friction between the slope and the prism is 0.32.

b Find whether the solid will slide or topple down the slope first and the angle at which this occurs.

20

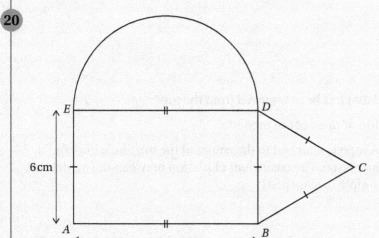

The diagram shows a uniform lamina $ABCDE$ formed by joining an equilateral triangle, a semicircle and a rectangle.

a Find the horizontal distance of the centre of mass from AE.

b Find the vertical distance of the centre of mass from AB.

The lamina is freely suspended from A and hangs in equilibrium.

c Find the angle that AB makes with the vertical.

Mathematics in life and work

You have designed a model. It is in the shape of a frustum of a pyramid (the portion of a pyramid that remains after the upper part has been removed) of uniform volume. The base of the pyramid measures 1 m by 1 m, the top measures 50 cm by 50 cm and the model is 1.15 m high.

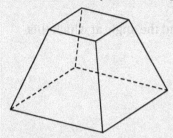

1   What is the height of the pyramid that has been removed from the top?

2   How far above the base is the centre of mass of the model?

3   The mass of the model is 125 kg. A rope is attached to the centre of the top and a horizontal force is applied that is gradually increased. The coefficient of friction between the model and the ground is 0.4. Will the model topple or slide first?

3 CIRCULAR MOTION

Mathematics in life and work

In this chapter, you will learn how to describe and analyse motion in horizontal and vertical circles. You will study the motion of conical pendulums and use the principle of the conservation of energy to predict what will happen when an object is moving in a vertical circle. This technique is important in a range of different careers – for example:

› If you were working in astronomy, you would be able to use of your knowledge of circular motion to investigate the orbits of moons around planets.

› If you were designing roads, you would use your knowledge of motion in a circle to help plan the size of circular bends and roundabouts and to decide what speed limits need to be put into place.

› If you were designing a theme park ride that moved in a circle, you would need to analyse the forces involved to ensure that riders will remain secure in their seats during the ride.

LEARNING OBJECTIVES

You will learn how to:

› understand the concept of angular speed for a particle moving in a circle, and use the relation $v = r\omega$

› understand that acceleration of a particle moving in a circle with constant speed is directed towards the centre of the circle, and use the formulae $a = r\omega^2$ and $\dfrac{v^2}{r}$

› solve problems that can be modelled by the motion of a particle moving in a horizontal circle with constant speed

› solve problems that can be modelled by the motion of a particle in a vertical circle without loss of energy.

LANGUAGE OF MATHEMATICS

Key words and phrases you will meet in this chapter:

› angular speed, centripetal force, conical pendulum, conservation of energy, period, tangential speed

PREREQUISITE KNOWLEDGE

You should already know how to:

> use radians as a measure of angle

> work out arc length by using $s = r\theta$

> understand Newton's laws of motion

> use the concepts of gravitational potential energy and kinetic energy

> use the principle of conservation of energy.

You should be able to complete the following questions correctly:

1 Convert 720° into radians.

2 How many radians would you rotate through if you turned around in a full circle five times?

3 What is the length of the arc in the diagram on the right? (Hint: Remember to convert the angle to radians.)

12 cm

4 A block of 5 kg is raised to 6 m. What is its change in gravitational potential energy?

5 A car of mass 3000 kg is travelling at a speed of $8 \, \text{m s}^{-1}$. What is its kinetic energy?

6 A rollercoaster car of mass 500 kg is moving at $15 \, \text{m s}^{-1}$ at a height of 10 m above the ground. What is the total amount of gravitational potential energy and kinetic energy?

3.1 Angular speed

In this section, you will learn how to calculate the angular speed of an object moving in a circle at a constant speed. Previously, you have calculated the speed of an object moving in a straight line by considering it as the rate at which distance is changing. For motion in a circle it is often easier to consider the rate at which the radius is rotating (i.e. the rate at which the angle is changing).

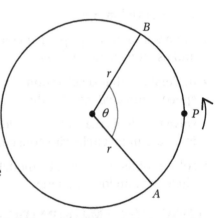

The diagram shows a particle P moving around the circumference of a circle with radius r and constant linear speed v. As the particle moves from point A to point B it turns through an angle of θ radians.

The speed of the particle is $v = \dfrac{s}{t} = \dfrac{r\theta}{t}$ as $s = r\theta$. • ——— s represents distance.

The **angular speed**, ω, is defined as $\dfrac{\theta}{t}$ and is • ——— ω is the Greek letter omega.
measured in rad s^{-1}.

Using this, the linear speed of the particle is given by $v = r\omega$. • ——— The linear speed is sometimes called the **tangential speed**.

Example 1

A particle is moving around a circle of radius 50 cm at a speed of $4\,\text{m}\,\text{s}^{-1}$. Find its angular speed.

Solution

$\omega = \dfrac{v}{r}$

$\quad = \dfrac{4}{0.5}$

$\quad = 8\,\text{rad}\,\text{s}^{-1}$

KEY INFORMATION

The angular speed of a particle, ω, is given by $\omega = \dfrac{\theta}{t}$.

The linear speed of a particle moving around a circle with radius r is given by $v = r\omega$.

The speed is given in metres per second so you need to convert the radius to m.

Example 2

A small ball is rolled around a smooth circular track. The ball moves around the track at a speed of $10\,\text{m}\,\text{s}^{-1}$ and takes 5 seconds to complete one full lap of the track. Find the diameter of the circular track.

Solution

Angular speed $\omega = \dfrac{\theta}{t}$

$\qquad \omega = \dfrac{2\pi}{5}$

$\qquad\quad = 1.2566\,\text{rad}\,\text{s}^{-1}$

One full lap is a turn of 2π radians

Since $v = r\omega$, $r = \dfrac{v}{\omega}$

$\qquad\quad = \dfrac{10}{1.2566}$

$\qquad\quad = 7.958\,\text{m}$

So the diameter of the track $= 2 \times 7.958 = 15.9\,\text{m}$ (3 s.f.)

Exercise 3.1A

1 A wheel is rotating at 60 r.p.m. (revolutions per minute). Find its angular speed, in radians per second.

2 Find the angular speed of the second hand on a clock, in radians per second.

3 A fly is sitting on the end of the second hand of a clock. If the second hand is 10 cm long, at what speed is the fly moving, in metres per second?

(MM) **4** A child is sitting on a roundabout, 1 m from the centre. The roundabout makes 15 turns per minute.

    a Find the angular speed of the child, in radians per second.

    b Find the linear speed of the child, in metres per second.

© **Communication**   (MM) **Mathematical modelling**   (PS) **Problem solving**

5 A ball is attached to the end of a light string of length 60 cm and is spun around in a horizontal circle. In 10 seconds it rotates in a circle five times. Find its angular and linear velocity.

 6 A child is running around a circular track of radius 30 m at a speed of 5 m s^{-1}.

    a Find the time taken to complete one lap of the track.

    b Find the child's angular velocity.

 7 Priya and Sanjay are running around a circular track at a constant speed of 4.5 m s^{-1}. Priya is running in lane 1, which is 42 m from the centre of the circular track, and Sanjay is running in lane 3, which is 44 m from the centre of the circular track.

    a Find the difference in the time taken for the runners to complete one lap of the track.

    b Find the difference in the angular velocity of the runners.

 8 A cyclist is racing around a circular track at a speed of 15 m s^{-1}. He can complete one lap in 35 s. Find the radius of the circular track.

 9 The radius of the Earth is approximately 6371 km. It rotates around the line connecting the North Pole and South Pole approximately once every 24 hours. What is the speed, in metres per second, of a person sitting on the equator?

3.2 Acceleration and forces for a particle moving in a circle

In Section 3.1, you studied the angular speed of a particle moving in a circle with constant speed. Although the speed is constant, the velocity of the particle cannot be constant because its direction is changing. In fact, at any point in time, the direction of the velocity is at a tangent to the circle. Since this velocity is changing, the particle must be accelerating.

It can be shown that this acceleration has magnitude $r\omega^2$ which is equivalent to $\dfrac{v^2}{r}$ acting towards the centre of the circle. As with linear motion, the units for acceleration are m s^{-2}.

> **KEY INFORMATION**
>
> The centripetal acceleration of a particle moving in a circle is directed towards the centre of the circle and can be calculated using the formulae
> $a = r\omega^2$ or $a = \dfrac{v^2}{r}$.

Example 3

A child is ice skating in a circle of radius 3 m at a speed of 20 km h^{-1}. Find the magnitude and direction of her acceleration.

Solution

To ensure consistency of units, convert the speed into metres per second.

$20 \, \text{km} \, \text{h}^{-1} = 20 \times 1000 \, \text{m} \, \text{h}^{-1}$

$\qquad = 20\,000 \, \text{m} \, \text{h}^{-1}$

$\qquad = 20\,000 \div (60 \times 60) \, \text{m} \, \text{s}^{-1}$

$\qquad = 5.56 \, \text{m} \, \text{s}^{-1}$

$\text{Acceleration} = \dfrac{v^2}{r}$

$\qquad\quad = \dfrac{5.56^2}{3}$

$\qquad\quad = 10.3 \, \text{m} \, \text{s}^{-2}$

So the acceleration is $10.3 \, \text{m} \, \text{s}^{-2}$ directed towards the centre of the circle.

The acceleration of a particle of mass m moving in a circle is towards the centre of the circle. By Newton's second law, the resultant force acting on the particle must also be acting towards the centre of the circle and have a magnitude of $mr\omega^2$ or $\dfrac{mv^2}{r}$. This is often called the **centripetal force**.

> **KEY INFORMATION**
>
> The resultant force acting on a particle of mass m moving in a circle can be calculated using the formulae
>
> $F = mr\omega^2$ or $F = \dfrac{mv^2}{r}$.
>
> This resultant force acts towards the centre of the circle.

Example 4

A car of mass 2000 kg is travelling around a circular bend of radius 100 m at a constant speed. The coefficient of friction between the car and the road is 0.7. Find the maximum linear speed at which the car can travel without moving in a sideways direction.

Solution

Consider the forces acting on the car by drawing a force diagram.

At the point of moving sideways, the frictional force $= 0.7 \times 2000g$

$\qquad\qquad\qquad = 14\,000 \, \text{N}.$

The force towards the centre of

the circle $= \dfrac{mv^2}{r}$

$\qquad\qquad = \dfrac{2000v^2}{100}$

At the point of sliding these forces will be equal:

$\dfrac{2000v^2}{100} = 14\,000$

$v^2 = 700$

$v = 26.5 \, \text{m s}^{-1}$

$= 26.5 \times 60 \times 60 \div 1000 \, \text{km h}^{-1}$

$= 95.4 \, \text{km h}^{-1}$

This means that the car can travel round the bend without moving in a sideways direction provided its speed is less than $95.4 \, \text{km h}^{-1}$.

Exercise 3.2A

1 Find the acceleration of a particle moving at a speed of $3 \, \text{m s}^{-1}$ on the edge of a spinning disc of radius 25 cm.

2 A particle is moving around in a horizontal circle with radius 3 m at a speed of $4.5 \, \text{m s}^{-1}$. Calculate the acceleration of the particle.

3 A car is driving on a section of road, which is a circular arc of radius 45 m, at a speed of $60 \, \text{km h}^{-1}$. Calculate the magnitude and direction of the car's acceleration.

4 A small toy of mass 30 g moves around in a horizontal circle of radius 20 cm at a constant speed of $4.8 \, \text{m s}^{-1}$. Find the magnitude and direction of the resultant force on the object.

(MM) 5 A disc is rotating in a horizontal circle with a constant angular speed of $3 \, \text{rad s}^{-1}$. A small block of mass 50 g is placed on the disc 25 cm from the centre.

   a Draw a force diagram for the block.

   b Find the magnitude of the friction force acting on the block.

(MM) 6 A car of mass 2000 kg is travelling around a bend which is a circular arc of radius 50 m. The greatest speed at which the car can travel around the bend without sliding is $50 \, \text{km h}^{-1}$. Find the coefficient of friction between the tyres of the car and the road.

(MM) 7 A motorbike of mass 600 kg is travelling around a circular bend of radius 42 m. The coefficient of friction between the motorbike and the road is 0.6. Find the maximum speed at which it can travel without sliding.

(MM) 8 A small toy of mass 100 g is placed on the edge of a circular turntable with radius 25 cm. The coefficient of friction between the toy and the turntable is 0.6. At what speed will the turntable need to be rotating for the toy to slide off the turntable?

(MM) 9 A car of mass 1500 kg is travelling around a circular bend of radius r in heavy rain. The coefficient of friction between the car and the road is 0.2. The driver finds that he can drive at a maximum speed of $45 \, \text{km h}^{-1}$ without sliding. What is the radius of the bend?

(PS) 10 A small object of mass m is placed x m from the centre of a turntable. The coefficient of friction between the particle and the turntable is μ. Show that the object will not slide if the angular speed, ω, satisfies $\omega^2 \leq \dfrac{10\mu}{x}$.

Mathematics in life and work: Group discussion

You are working for a large theme park and are responsible for ensuring customer safety on all rides. A new go-kart track is due to open and you have been asked to investigate what speed limits have to be put in place to ensure that the go-karts can move safely around the track without sliding. Your first task it to work out the coefficient of friction between the track and the go-kart's tyres in different weather conditions.

1 How could you find the coefficient of friction between the tyres and the track?

You find that the coefficient of friction between the tyres and a dry track is 0.7 and between the tyres and a wet track is 0.4. On the track there are three different circular bends, none of which is banked (sideways sloped). The first bend has a radius of 25 m, the second has a radius of 30 m and the third has a radius of 40 m. The average mass of one of the go-karts with a passenger is 200 kg.

2 What would you suggest as the maximum speed limit on a dry day?

3 How would you change your suggestion if the track is wet?

4 What impact would having a banked track have on the speed limit?

5 One of the park managers is concerned that you have used the average mass of the go-kart. He queries whether you need to put a weight limit on the ride. How would you respond to the manager?

3.3 Objects moving in horizontal circles

In this section, you will resolve forces to solve problems about motion in a horizontal circle.

A **conical pendulum** consists of a particle of mass m attached to a light inextensible string of length l. As the particle moves in a horizontal circle, the string traces the outline of a cone; hence the name 'conical pendulum'.

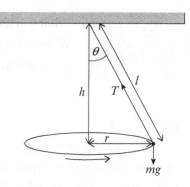

You know that the particle is accelerating with magnitude $r\omega^2$ towards the centre of the circular base. This gives a resultant force, acting towards the centre of the circle, of magnitude $mr\omega^2$.

The forces acting on the particle are its weight, mg, and the tension in the string, T.

Resolving these vertically:

$T \cos \theta = mg$

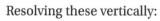

$$T = \frac{mg}{\cos\theta}$$

> The acceleration of the particle has no vertical component.

Resolving these horizontally, by Newton's second law:

$T \sin \theta = ma$

$T \sin \theta = mr\omega^2$

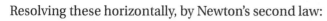

$$T = \frac{mr\omega^2}{\sin\theta}$$

Equating the above two equations:

$$\frac{mg}{\cos\theta} = \frac{mr\omega^2}{\sin\theta}$$

From the above diagram you can see that $\sin\theta = \frac{r}{l}$.

$$\frac{g}{\cos\theta} = \frac{r\omega^2}{\frac{r}{l}}$$

$$\omega^2 = \frac{g}{l\cos\theta}$$

Using this relationship, you can solve a variety of problems about a conical pendulum.

You can also use this result to help you to calculate the period of the pendulum.

$$t = \frac{2\pi}{\omega}$$

$$t = \frac{2\pi}{\sqrt{\frac{g}{l\cos\theta}}}$$

$$t = 2\pi\sqrt{\frac{l\cos\theta}{g}}$$

> The **period** of a pendulum is the time taken to complete one full circle.

Stop and think

What will happen if the length of the pendulum is increased?

What will happen if the radius of the circle is increased?

What will happen if the mass of the object is increased?

Example 5

A particle of mass 0.5 kg is attached to the end of a light inextensible string of length 0.8 m. The other end is fixed in place so that the particle moves in a horizontal circle at a constant speed. The string makes an angle θ with the vertical, where $\cos\theta = 0.8$.

a Find the radius of the circle made by the pendulum.

b Find the angular speed of the particle.

c Find the tension in the string.

Solution

This problem can be represented in a force diagram.

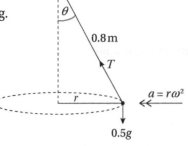

a From the diagram,

$\sin\theta = \frac{r}{0.8}$, so

$r = 0.8\sin\theta$.

Since you are given the value of cos θ, you can use a right-angled triangle to evaluate sin cos θ.

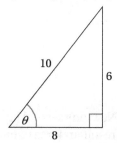

$\cos\theta = 0.8$

$= \frac{8}{10}$

so $\sin\theta = \frac{6}{10}$

$= 0.6$

$\sin\theta = 0.6$, so $r = 0.8 \times 0.6 = 0.48\,\text{m}$

b For a conical pendulum, $\omega^2 = \dfrac{g}{l\cos\theta}$, so:

$$\omega = \sqrt{\frac{g}{l\cos\theta}}$$

$$= \sqrt{\frac{10}{0.8 \times 0.8}}$$

$$= 3.95\,\text{rad}\,\text{s}^{-1}\ (3\ \text{s.f.})$$

c Resolving vertically:

$T\cos\theta = 0.5g$

$$T = \frac{0.5 \times 10}{0.8}$$

$$= 6.25\,\text{N}\ (3\ \text{s.f.})$$

Understanding how to solve problems for a conical pendulum can help in slightly different scenarios, as you will see in the next two examples.

Example 6

An object P of mass 5 kg is attached to two inextensible strings of lengths 5 m and 13 m. The strings are attached on a rotating vertical pole at points A and B so that the particle can move in a horizontal circle. The time taken to complete one full revolution is 2 seconds.

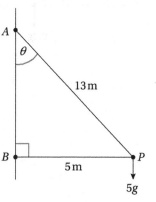

Solution

Draw a force diagram for the scenario.

Before you continue, it will be helpful to know the angular velocity of the particle.

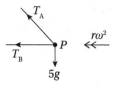

$\omega = \dfrac{\theta}{t} = \dfrac{2\pi}{2} = \pi\ \text{rad}\,\text{s}^{-1}$

Resolving vertically:

$T_A \cos \theta = mg$

$T_A \times \frac{12}{13} = 5 \times 10$

$T_A = 54.2\,\text{N}$

Using Newton's second law
(towards the centre):

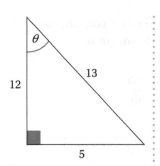

$T_A \sin \theta + T_B = mr\omega^2$

$T_B = mr\omega^2 - T_A \sin \theta$

$\quad = 5 \times 5 \times \pi^2 - 54.2 \times \frac{5}{13}$

$\quad = 226\,\text{N (3 s.f.)}$

> Using Newton's second law with a resultant force equal to $mr\omega^2$.

So the tension in string A is 54.2 N and the tension in string B is 226 N.

Example 7

A small ball of mass 200 grams moves in a horizontal circle of radius r on the inside surface of a smooth cone. The angle at the vertex of the cone is 50° as shown in the diagram. Find the radius of the circle if the ball is travelling at a speed of $6\,\text{m s}^{-1}$.

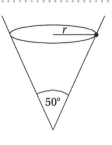

Solution

Draw a force diagram for the ball.

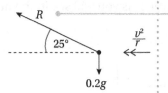

Resolving vertically:

$R \sin 25° = 0.2g$

$R = 4.73\,\text{N}$

> The normal reaction force, R, is acting perpendicular to the edge of the cone.

Using Newton's second law:

$R \cos 25° = \dfrac{mv^2}{r}$

$r = \dfrac{mv^2}{R \cos 25°}$

$\quad = \dfrac{0.2 \times 6^2}{4.73 \cos 25°}$

$\quad = 1.68\,\text{m}$

The radius of the cone is 1.68 m (3 s.f.).

Exercise 3.3A

 1 A small object of mass 2.5 kg is tied to the end of a piece of string of length 0.5 m to make a conical pendulum. The angle between the string and the vertical is 30°. Calculate:

    a the radius of the circular path

    b the period of the pendulum

    c the tension in the string.

 2 A piece of putty of mass 0.4 kg is attached to the end of a piece of string of length 55 cm to create a conical pendulum which rotates in a circle of radius 35 cm. Calculate:

    a the angle between the string and the vertical

    b the tension in the string

    c the acceleration of the piece of putty

    d the speed of the piece of putty.

 3 A particle of mass 600 g is attached to one end of a light inextensible string of length 90 cm. The other end of the string is attached to a fixed point. The particle rotates in a horizontal circle, the centre of which is 0.6 m directly below the fixed point. Calculate:

    a the tension in the string

    b the angular speed of the particle.

4 A small ball is moving at a speed of 5 m s^{-1} in a horizontal circle of radius r on the inside surface of a smooth cone. Find the radius r, of the horizontal circle, given that the mass of the ball is 350 g and the angle at the centre of the cone is 40°.

 5 A smooth hemispherical bowl has a radius of 0.3 m. A small ball of mass 80 g is rotating in a horizontal circle on the inside of the bowl, with an angular speed of 10 rad s^{-1}. Find the distance between the centre of the circle and the base of the bowl.

 6 A ball of mass 300 g is moving in a horizontal circle of radius 40 cm on the inside surface of a smooth hemisphere of radius 60 cm. Find the angular speed of the ball.

7 A conical pendulum consists of an inextensible string fixed at a point, with a small object of mass m attached to the other end. The small object follows a circular path with a constant angular velocity of ω. Show that $\omega = \sqrt{\dfrac{g}{h}}$, where h is the distance between the fixed point and the centre of the circle.

 8 A small ball is made to rotate in a horizontal circle of radius r on the inside of a smooth cone with an angle of 2θ at the base vertex. It is travelling with a constant angular velocity of ω. Show that

$$\omega = \sqrt{\frac{g}{r\tan\theta}} \, .$$

Mathematics in life and work: Group discussion

As part of your job at the theme park, you are responsible for showing groups of students around the park. You are demonstrating the mechanics of the suspended chairs carousel, which consists of a number of chairs suspended from chains that swing out as the centre rotates faster and faster.

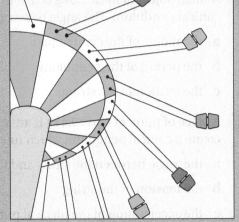

As part of the tour, students are asked to work out the linear speed of people on the ride. Once the ride has reached its maximum speed, they measure the angle between the chains and the vertical, as well as the radius of the circle they make.

The students find that, for those in the outer chairs, the angle between the chains and the vertical is 32° and the radius of the circle made by the chairs is 5.2 m. They measure the length of the chain to be 10 m.

1   Assuming that the average rider has a mass of 40 kg, find the linear speed at which the riders will move.

One of the students is concerned that people of different masses would ride in circles of different radii, which could result in empty seats and heavier passengers colliding.

2   Is the student right to be concerned? Explain why.

3.4 Motion in a vertical circle

In Section 3.3, the speed of the object remained constant as it moved around the circle. When the path of an object follows a vertical circle, the speed, v m s^{-1}, of the object changes as it moves around the circle. The speed is at its maximum at the bottom of the circle and at its minimum at the top of the circle.

This change in speed means that the object's acceleration must have two components – one acting towards the centre of the circle, with magnitude $\frac{v^2}{r}$, and the other acting along the tangent of the circle, with a magnitude equal to the rate of change of speed, $\frac{dv}{dt}$.

> **KEY INFORMATION**
>
> If an object is moving in a circle with variable speed v m s^{-1}, its acceleration will have two components:
>
> › $\frac{v^2}{r}$ acting towards the centre of the circle
>
> › $\frac{dv}{dt}$ acting along the tangent of the circle.

Stop and think   What does this mean for the resultant force acting on the object? Where in the circle will this force be at its maximum and where will it be at its minimum?

If no energy is lost as the object moves around the circle, then the principle of **conservation of energy** will help to analyse the motion in detail. In this context, this principle states that the sum of the potential and kinetic energy of a system remains constant if no energy is lost. The next example shows how this can be used.

Stop and think   Can you think of examples of how energy could be lost from a system?

Example 8

A small object of mass 0.5 kg is attached to the end of a light rod AB of length 60 cm. The rod is free to rotate around B. The rod is held horizontally and then released from rest.

a Find the speed of the object at the lowest point of the circle.

b Find the tension in the rod at this point.

Solution

The information given in the question can be represented by the following diagram.

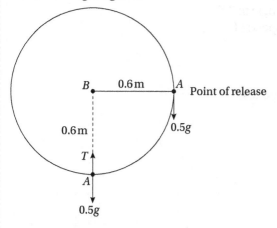

To make the calculations easier, let the lowest part of the circle have a height of 0.

a At the point of release, the object will have only potential energy due to it being at rest.

$$PE = mgh = 0.5 \times 10 \times 0.6 = 3\,J$$

No energy is lost so, by the principle of the conservation of energy, all of this potential energy will be converted into kinetic energy at the lowest point of the circle. ●───────

> It will have no potential energy because it has a height of 0 at this point.

$$KE = \frac{1}{2}mv^2 = 3$$

$$\frac{1}{2} \times 0.5 \times v^2 = 3$$

$$v^2 = 12$$

$$v = 3.46\,\text{m s}^{-1}$$

So the speed at the lowest point of the circle is $3.46\,\text{m s}^{-1}$.

b At the lowest point in the circle, the resultant force is

equal to $\frac{mv^2}{r}$.

Applying Newton's second law at the lowest point:

Resultant force = $T - 0.5g$

$T - 0.5g = \frac{mv^2}{r}$

$T - 0.5g = \frac{0.5 \times 3.46^2}{0.6}$

$T = 15.0\,\text{N}$ (3 s.f.)

Example 9

A father and daughter are playing on a swing. The swing has a rope of length 2.5 m. The father raises the swing at an angle of 30° to the vertical and pushes his daughter with an initial speed of $2\,\text{ms}^{-1}$. His daughter has a mass of 25 kg. Find:

a the maximum speed of his daughter

b the angle from the vertical for which the swing reaches its maximum height.

Solution

This question can be summarised by this diagram.

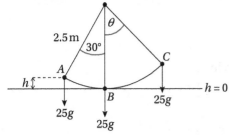

a His daughter is released from A with a speed of $2\,\text{ms}^{-1}$. She reaches her maximum speed at B and reaches her maximum height at C.

To find her potential energy at A you need to know the height, which you can find using trigonometry.

First you need to find x, and then the height, $h = 2.5 - x$.

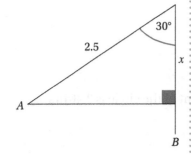

$\cos 30° = \frac{x}{2.5}$

$x = 2.5\cos 30°$

$= 2.165\,\text{m}$

So $h = 2.5 - 2.165$

$= 0.335\,\text{m}$

At A, total energy $= \frac{1}{2}mv^2 + mgh$

$$= 0.5 \times 25 \times 2^2 + 25 \times 10 \times 0.335$$

$$= 133.75\,\text{J}$$

No energy is lost so, by the principle of the conservation of energy, all of this potential energy will be converted into kinetic energy at the lowest point of the circle.

KE at B $= \frac{1}{2}mv^2 = 133.75$

$$\frac{1}{2} \times 25 \times v^2 = 133.75$$

$$v^2 = 10.7$$

$$v = 3.27\,\text{m s}^{-1}$$

The speed at the lowest point of the circle is $3.27\,\text{m s}^{-1}$.

b To find the angle made by the rope when the swing is at its maximum height, first find the maximum height. At the maximum height the girl's speed will be zero so she will have no kinetic energy.

Total energy at maximum height $= mgh = 133.75$

$$25 \times 10 \times h = 133.75$$

$$h = 0.535\,\text{m}$$

This gives the following triangle:

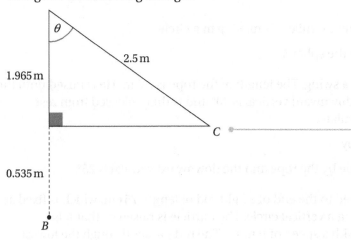

These sketches are to help with the calculations and are not to scale.

$$\cos\theta = \frac{1.965}{2.5}$$

$$\theta = 38.2° \text{ (to 1 d.p.)}$$

The angle made by the rope and the vertical at the maximum height is 38.2°.

Exercise 3.4A

1 An object of mass 2 kg is attached to the end of a rope of length 1.5 m. The other end of the rope is attached to a fixed point. The object is raised so that the string is taut and horizontal, and released so that it moves in a vertical circle with an initial speed of 3 m s^{-1}. Assuming that the height of the lowest part of the circle is 0 m, calculate:

a the total energy of the object at the point of release

b the speed of the object at its lowest point.

2 A particle of mass 300 g is fixed to one end of a rod of length 70 cm and moves in a vertical circle about a fixed point. At the highest point of the circle the particle has a speed of 1.5 m s^{-1}. Find the speed at the lowest point of the circle.

3 A child of mass 30 kg is sitting on a swing. The swing is attached to the frame by an inextensible rope of length 2.4 m. The child's mother pulls the swing back until the rope makes an angle of 30° with the vertical and releases it from rest. Calculate, stating any assumptions you make:

a the height the swing is released from

b the total energy of the child at the point of release

c the speed at the lowest point of the circle.

4 A light inextensible rod of length 2 m is attached to a fixed point. A small sphere of mass 3 kg is attached to the other end. The sphere is raised so that it is in a horizontal position and is then released from rest. Assuming that the lowest point of the circle has zero potential energy, calculate:

a the maximum speed of the sphere while it is moving in a circle

b the maximum force acting on the sphere.

5 A boy of mass 45 kg is playing on a swing. The length of the rope is 2.5 m. He is raised until the angle between the rope and the downward vertical is 40° and is then released from rest. Showing all of your working, calculate:

a the maximum speed of the boy

b his speed when the angle made by the rope and the downward vertical is 25°.

6 A particle of mass 0.5 kg is attached to the end of a light rod of length 75 cm, which is fixed at the other end to allow it to move in a vertical circle. The particle is raised so that it is horizontal and is then released with a speed of u m s^{-1}. The rod passes through the lowest point of the circle and comes to rest when the angle made with the downward vertical is 110°.

a Find the height of the particle above the base of the circle when it comes to rest.

b Find the speed with which the particle was released.

7 A small sphere of mass m is attached to a light rod of length l, which is attached to a fixed point that allows it to move in a vertical circle. The speed, v, of the sphere at the top of the circle is $\frac{1}{4}$ of the speed at the bottom of the circle.

a Show that the length of the rod is given by $l = \dfrac{15v^2}{4g}$.

b Find expressions for the maximum and minimum centripetal forces acting on the sphere as it moves around the circle.

 8 A particle of mass m is attached to the end of a light inextensible string of length r, which is attached to a fixed point A. The particle starts at the lowest point of the circle and is sent moving in a circle with a speed of $\sqrt{3gr}$. Show that the speed of the particle, v, at any point of the circle is given by $v = \sqrt{gr(1 - 2\cos\theta)}$, where θ is the angle made with the upward vertical.

3.5 The conditions for complete circular motion

In this section, you will investigate the conditions needed for a particle to move in a complete circle – and what will happen if it does not move in a complete circle.

You will look at two main cases:

1 When the object is fixed to move around the circumference of a circle – for example, being attached to a light rod or wire.

2 When the object is able to move away from the circumference of a circle – for example, being attached to a light inextensible rope.

In the first case, what happens depends on the speed of the object. If the speed of the object is greater than or equal to zero at the top of the circle then it will move in a complete circle. Otherwise, it will instantaneously come to rest at a point before the top of the circle, and then it will oscillate around the bottom of the circle.

> **KEY INFORMATION**
>
> If an object is fixed to move around in a circle, then it will either:
>
> ❭ move around in a complete circle if the speed at the top is greater than zero, or
>
> ❭ instantaneously come to rest before it reaches the top and then oscillate around the bottom of the circle.

Example 10

A small bead of mass 25 g is attached to a 30 cm length of light rigid wire. The wire is fixed at one end to allow it to move around in a circle. The bead is released from A with a speed of $3\,\text{m s}^{-1}$, as shown in the diagram. Describe what will happen to the bead.

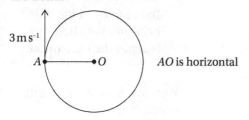

AO is horizontal

Solution

Assume that the base of the circle has height 0 m (or zero potential energy).

Energy of bead at $A = \frac{1}{2}mv^2 + mgh$

$$= 0.5 \times 0.025 \times 3^2 + 0.025 \times 10 \times 0.3$$

$$= 0.1875\,\text{J}$$

If the bead is to reach the top of the circle, then, by conservation of energy:

Energy at top $= \frac{1}{2}mv^2 + mgh = 0.1875\,\text{J}$

So $0.5 \times 0.025 \times v^2 + 0.025 \times 10 \times 0.6 = 0.1875$

$$0.0125v^2 = 0.0375$$

$$v = 1.73\,\text{m s}^{-1}$$

Since $v > 0$, the bead is going to move in a complete circle.

Stop and think

In Example 10, you found that the bead would move around in a complete circle. What is the lowest possible speed of release for this to happen?

Now consider what happens when the object is free to move away from the circle – for example, if it is attached to a string. In this case, the outcome depends mainly on the tension in the string. There are three main cases.

Case 1:

If the tension at the top of the circle is greater than or equal to zero (i.e. the string is taut), the object will move in a complete circle.

Case 2:

If the tension drops to zero above the horizontal but before the top, the particle will move as a projectile.

Case 3:

If the velocity of the object reaches zero before it reaches the horizontal, it will oscillate about the bottom of the circle.

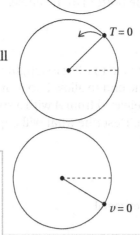

KEY INFORMATION

If an object is free to move away from the circle then:

> it will move in a complete circle if the tension at the top of the circle is greater than or equal to zero

> it will move as a projectile if the tension becomes zero between the horizontal and the top of the circle

> it will oscillate around the bottom of the circle if the velocity of the object becomes zero before it reaches the horizontal.

You saw this happening in Example 9.

Example 11

An object of mass 20 g is attached to the end of a light inextensible string of length 50 cm. The other end of the string is attached to a fixed point which allows the object to move around in a circle. The particle is released downwards at a speed of $2\,\mathrm{m\,s^{-1}}$ from a horizontal position. Describe what will happen to the object.

Solution

The diagram shows the object when it is θ above the vertical, where v is the velocity at any point on the circle and T is the tension in the string.

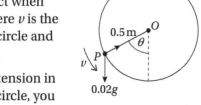

If you had a formula for the tension in the rope at any point of the circle, you could then work out when this tension was zero and hence see what is going to happen.

Start by finding the total energy in the system.

The particle is released from a horizontal position so its height will be 0.5 m.

So total energy $= \frac{1}{2}mu^2 + mgh$

$$= 0.5 \times 0.02 \times 2^2 + 0.02 \times 10 \times 0.5$$

$$= 0.14\,\mathrm{J}$$

At any point P, the particle can have both kinetic and potential energy.

Kinetic energy $= \frac{1}{2}mv^2 = 0.5 \times 0.02v^2 = 0.01v^2\,\mathrm{J}$

Potential energy $= mgh$

To find the height at any point P, use trigonometry.

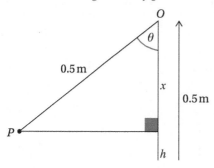

$\cos\theta = \dfrac{x}{0.5}$

$x = 0.5\cos\theta$

So $h = 0.5 - 0.5\cos\theta = 0.5(1 - \cos\theta)$

So potential energy at point $P = 0.02 \times 10 \times 0.5(1 - \cos \theta)$

$$= 0.1(1 - \cos \theta)$$

So total energy at point $P = 0.01v^2 + 0.1(1 - \cos \theta) = 0.14$ •————— Because energy is conserved.

This means at any point, $v^2 = \dfrac{0.14 - 0.1(1 - \cos\theta)}{0.01}$

$$= 14 - 10(1 - \cos \theta)$$

$$= 4 + 10 \cos \theta$$

By Newton's second law parallel to OP:

$$T - 0.02g \cos \theta = \frac{mv^2}{r}$$

$$T - 0.02g \cos \theta = \frac{0.02(4 + 10\cos\theta)}{0.5}$$

$$50T - g \cos \theta = \frac{(4 + 10\cos\theta)}{0.5}$$

$$50T - 10 \cos \theta = 8 + 20 \cos \theta$$

$$50T = 30 \cos \theta + 8$$

When the tension is zero:

$$30 \cos \theta + 8 = 0$$

$$\cos \theta = -\frac{8}{30}$$

$$\theta = 105.5°$$

This tells you that the object will leave the circle when $\theta = 106°$ and will follow the path of a projectile.

Stop and think   At the point when the object begins to act as a projectile, what is the angle of projection from the horizontal? What will its speed be at this point?

Sometimes the particle will be moving in a circle on the inside or outside of a spherical surface. In this case, the normal reaction of the surface on the particle is a force acting towards the centre of the circle. This can be seen in the next example.

Example 12

A toy car of mass 500 g travels along the smooth track of a toy rollercoaster, which loops around in a circle of radius 40 cm. The toy car can move freely along the track but cannot leave it. The car enters the loop at its lowest point with a speed of $5 \, \text{m s}^{-1}$.

a   Find the speed at the top of the loop.

b   Find the force acting on the car at the top of the loop.

Solution

a If you assume that the lowest point of the loop is at $h = 0$ m, then the car will only have kinetic energy at this point.

$$\text{Total energy on entry to the loop} = \frac{1}{2}mv^2$$

$$= 0.5 \times 0.5 \times 5^2$$

$$= 6.25 \text{ J}$$

The height of the top of the circle is 0.8 m so the car will have both kinetic and potential energy at this point.

$$\text{Total energy at top of loop} = \frac{1}{2}mv^2 + mgh$$

$$= 0.5 \times 0.5 \times v^2 + 0.5 \times 10 \times 0.8$$

$$= 0.25v^2 + 4$$

By the principle of conservation of energy:

$$0.25v^2 + 4 = 6.25$$

$$v^2 = 9$$

$$v = 3 \text{ m s}^{-1}$$

So at the top of the loop the car has a speed of 3 m s^{-1}.

b

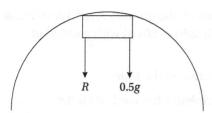

$$R \qquad 0.5g$$

By Newton's second law:

$$R + 0.5g = \frac{mv^2}{r}$$

$$R + 5 = \frac{0.5 \times 3^2}{0.4}$$

$$R = 6.25 \text{ N}$$

So the force acting on the car at the top of the loop is 6.25 N downwards.

Stop and think   What would happen to the car if it was not fixed to the track?

Mathematics in life and work: Group discussion

A new rollercoaster is to be built. You have been asked to look over the final plans.

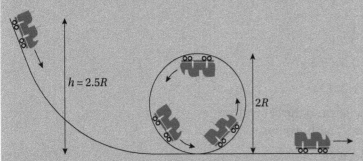

One of the designers has researched online and found that the minimum height of the start of the track needs to be at least 2.5 times the radius of the loop, as shown in the diagram. Another designer doubts this claim.

Show that the first designer is correct.

Exercise 3.5A

 1 An object of mass 2 kg is swung in a vertical circle with a radius of 0.75 m. What is the minimum speed of the object at the bottom of the loop for the object to remain in circular motion?

 2 A smooth cylindrical pipe of radius 60 cm is lying on its side on the ground. A marble of mass 100 g is placed on the inside of the pipe. It is released at a height of 60 cm with a speed of 5 m s^{-1}.

    a What is the speed of the marble when it reaches the bottom of the pipe?

    b What condition must be satisfied for the marble to complete a full vertical circle?

    c Will the marble complete a full vertical circle inside the pipe? You must show calculations to back up your answer.

 3 A small sphere of mass 250 g is attached to the end of a light rod of length 1 m. The rod is attached to a fixed point, allowing it to move in a vertical circle. The sphere leaves the bottom of the circle with a speed of 4 m s^{-1}.

    a Calculate the height reached by the sphere before it comes to rest.

    b Calculate the angle made with the downward vertical when the sphere comes to rest.

 4 A particle of mass 100 g is attached to the end of a piece of inextensible string of length 50 cm. The string is attached to a fixed point, allowing the particle to move in a vertical circle. The particle is released from the lowest point of the circle with a speed of u.

    a Find the range of values of u that will result in the particle oscillating around its starting point.

    b Find the range of values of u that will result in the particle completing a full circle.

PS **5** An object of mass 40 g is attached to the end of a light inextensible string of length 60 cm. The other end of the string is attached to a fixed point, allowing the object to move around in a vertical circle. The particle is released downwards at a speed of 2.3 m s⁻¹ from a horizontal position. Describe what will happen to the object.

MM **6** A toy train of mass 0.4 kg travels along a smooth track that contains a circular loop of radius 50 cm. The toy train can move freely along the track but cannot leave it. The train enters the loop at its lowest point with a speed of 6 m s⁻¹.

    a Find the speed of the train at the top of the loop.

    b Find the resultant force acting on the train at the top of the loop.

MM **7** A smooth hemispherical bowl of radius 0.75 m is placed rim down on a table. A small sphere of mass 0.15 kg is placed on the highest part of the bowl and released with a speed of 2.3 m s⁻¹. The angle θ is the angle between the upward vertical and the line joining the position of the ball and the centre of the base of the bowl.

    a Find the total energy of the ball in terms of θ and v.

    b The ball will lose contact with the bowl when the normal reaction force on the ball is 0 N. Find the angle θ at which the ball loses contact with the bowl.

C **8** A ball of 3 kg is placed on the top of a smooth circular hemisphere with radius 2 m and released from rest. Describe what happens once the ball is released.

MM **9** A particle of mass m is suspended from a light inextensible string of length r. The string is attached to a fixed point, allowing the particle to rotate in a vertical circle. The particle passes the lowest part of the circle with a speed of u.

    a Assuming that the string remains taut, show that the speed v, at any position in the circle, is given by $v^2 = u^2 - 2gr(1 - \cos\theta)$, where θ is the angle the string makes with the downwards vertical.

    b Find an expression for the tension in the string at any given point.

SUMMARY OF KEY POINTS

➤ Angular speed of a particle, ω, is given by $\omega = \frac{\theta}{t}$ where θ is the angle though which the particle moves in t seconds. It is measured in rad s^{-1}.

➤ The linear speed of a particle moving around a circle with radius r is given by $v = r\omega$.

➤ The acceleration of a particle moving in a circle is directed towards the centre of the circle and can be calculated using the formula $a = r\omega^2$ or $a = \frac{v^2}{r}$.

➤ By Newton's second law, the resultant force acting on a particle of mass m moving in a circle can be calculated using the formula $F = mr\omega^2$ or $F = \frac{mv^2}{r}$. This resultant force acts towards the centre of the circle.

➤ A conical pendulum consists of a particle of mass m attached to a light inextensible string of length l moving in a horizontal circle.

➤ When the path of an object follows a vertical circle, the speed is at its maximum at the bottom of the circle and at its minimum at the top of the circle.

➤ The principle of conservation of energy states that the sum of potential and kinetic energy of a system remains constant if no energy is lost. This principle helps to calculate the speed of an object moving in a vertical circle at different points on its path.

➤ If an object is fixed to move around in a circle, then it will either:

   ➤ move around in a complete circle if the speed at the top is greater than zero, or

   ➤ instantaneously come to rest before it reaches the top and then oscillate around the bottom of the circle.

➤ If an object is free to move away from the circle, then:

   ➤ it will move in a complete circle if the tension at the top of the circle is greater than or equal to zero

   ➤ it will move as a projectile if the tension becomes zero between the horizontal and the top of the circle

   ➤ it will oscillate around the bottom of the circle if the velocity of the object becomes zero before it reaches the horizontal.

EXAM-STYLE QUESTIONS

1 One end of a light inextensible string of length 1.5 m is attached to a fixed point A and the other end of the string is attached to a particle P. The particle P moves at a constant speed in a horizontal circle of radius 50 cm, which has its centre O vertically below A. The particle P has a mass of 500 g.

a Find the speed of P.

b Find the period of the pendulum.

2 One end of a light inextensible string of length 1.3 m is attached to a fixed point *A* and the other end of the string is attached to a small sphere of mass 1.5 kg. The sphere moves at a constant speed in a horizontal circle of radius 60 cm, which has its centre *O* vertically below *A*. The sphere has a mass of 1.5 kg.

  a Find the angle between the string and the downward vertical.

  b Calculate the tension in the string.

3 A particle *P* of mass 20 kg is attached to one end of a light inextensible string of length 1 m. The other end of the string is attached to a fixed point *O*. When *P* is hanging at rest vertically below *O*, it is projected horizontally. In the subsequent motion, *P* completes a vertical circle. The speed of *P* when it is at its highest point is $2 \, \mathrm{m \, s^{-1}}$.

  a Find the maximum speed of *P*.

  b In the case when *P* is at its maximum height, find the tension in the string.

4 One end of a light inextensible string of length 5 m is attached to a fixed point *A* and the other end of the string is attached to a particle *P*. The particle *P* moves at a constant speed in a horizontal circle of radius 2.6 m, which has its centre *O* vertically below *A*. The particle *P* has a mass of 50 kg.

  a Calculate the speed of *P*.

  b Find the tension in the string.

5 A car of mass 2500 kg is driving around a circular bend of radius 80 m.

  a When the road is dry, the coefficient of friction between the car and the road is 0.7. Calculate the speed of the car when it is on the point of sliding.

  b When the road is wet, the coefficient of friction between the car and the road is μ. The speed of the car is $45 \, \mathrm{km \, h^{-1}}$ when it is on the point of sliding. Find the value of μ.

6 A small ball *B* of mass 200 g is placed on a horizontal turntable that rotates at a constant speed of 90 revolutions per minute. The coefficient of friction between *B* and the turntable is 0.4 and *B* is at a distance *d* from the centre of the turntable.

  a Calculate the angular speed of *B*, in radians per second.

  b Find the value of *d* when *B* is on the point of sliding.

7 A particle of mass 3 kg is attached to one end of a light inextensible string of length 1.5 m. The other end is fixed in place and the particle moves in a horizontal circle at a constant speed. The string makes an angle θ with the vertical where $\sin \theta = 0.28$.

  a Find the radius of the circle made by the particle.

  b Find the angular speed of the particle.

  c Find the tension in the string.

8 A rough turntable rotates about its centre *O* in a horizontal plane with a constant angular speed of 60 revolutions per minute. A small block *B* of mass 300 g is on the turntable and *OB* = 10 cm. The coefficient of friction between *B* and the turntable is μ. Given that *B* does not slide, find the smallest possible value of μ.

MM **9** One end of a light inextensible string of length $5x$ m is attached to a fixed point C. The string passes through a smooth bead B of mass $2m$ kg and the other end of the string is attached to a fixed point D vertically below C. The bead B moves with constant speed in a horizontal circle with centre D and radius $3x$ m.

a Find the tension in the string.

b Find the speed of the bead in terms of x.

MM **10** A particle P of mass 50 g is attached to one end of a light inextensible string of length 4 m. The other end of the string is attached to a fixed point O. Initially, OP makes an angle of 50° with the downward vertical and the string is taut. P is projected with a speed of $3\,\text{m s}^{-1}$ and follows the downward arc of a vertical circle with centre O.

a Find the maximum speed of P.

b Find the maximum height that P reaches above its starting position.

PS **11** A particle P of mass 30 g is attached to one end of a light inextensible string of length 60 cm.
C The other end of the string is attached to a fixed point O. Initially, OP is horizontal and the string is taut. P is projected downwards at a speed of $3\,\text{m s}^{-1}$ such that P moves in a vertical circular arc with centre O. When OP makes an angle θ below the horizontal, P is travelling at speed $v\,\text{m s}^{-1}$.

a Show that $v^2 = 9 + 12\sin\theta$.

b Find the exact value of θ when the string becomes slack.

C **12** A small sphere, S, of mass m kg is attached to one end of a light inextensible string of length x m. The other end is attached to a fixed point O. S is projected from its lowest point with a speed of $u\,\text{m s}^{-1}$, where $u^2 = 30x$, and follows the arc of a vertical circle with centre O. When the string is taut and OS makes an angle θ with the downward vertical, the tension in the string is T.

a Find an expression for T in terms of g, x and θ.

b Show that, when the string becomes slack, the speed of S is given by $v^2 = \dfrac{10}{3}x$.

MM **13** A smooth, solid hemisphere of radius 70 cm and centre O is fixed with its flat face on a horizontal surface. A small ball of mass 400 g is projected horizontally, with a speed of $u\,\text{m s}^{-1}$, from the point A at the top of the hemisphere. The small ball leaves the sphere at point B, which is 50 cm above the horizontal surface. The angle between OA and OB is θ.

a Find the value of $\cos\theta$.

b Find the speed with which the particle leaves the hemisphere.

c Find the value of u.

C **14** A smooth, hollow cone is placed vertex down with its circular face in a horizontal position.

The radius of the cone is $\dfrac{3}{4}$ of its height. A particle of mass m moves in a horizontal circle at a

constant angular speed of $\sqrt{\dfrac{80}{3r}}\,\text{rad s}^{-1}$. Find the vertical height of the cone in terms of r.

15 A particle P of mass m kg is attached to one end of a light inextensible string of length 0.5 m. The other end of the string is attached to a fixed point O. Initially, OP is horizontal and the string is taut. P is projected downwards with a speed of u m s^{-1} and follows the arc of a vertical circle with centre O.

a In the case where $u = 0$ m s^{-1}, find the maximum angular speed of P.

b In the case where P moves in a complete vertical circle, find the minimum value of u.

c In the case where P moves in a complete vertical circle and the maximum speed is

double the minimum speed, show that $u = \sqrt{\dfrac{5g}{3}}$.

16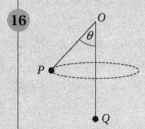

The diagram shows a particle P of mass 4 kg and a particle Q of mass 8 kg. P is attached to one end of a light inextensible string and Q is attached to the other end. The string passes through a smooth ring fixed at a point O, which is vertically above Q. Q hangs at rest and P moves in a horizontal circle of radius 2 m at a constant speed of v m s^{-1}. The angle between OP and the downward vertical is θ.

a Find the value of $\cos \theta$.

b Show that $v^2 = 2\sqrt{3}g$.

PS **17** A small sphere S of mass m kg is attached to one end of a light rod of length 0.75 m. The other end is attached to a fixed point O, allowing S to move in a vertical circle. When S is below O, it has a speed of v m s^{-1}. When S vertically above O, it has a speed of $\sqrt{v}$ m s^{-1}. Find the maximum speed of S.

18 A smooth, hollow hemisphere is placed with its curved surface downwards and its circular face, centre O, in a horizontal position. The radius of the hemisphere is r m. A particle P of mass m moves in a horizontal circle 1.5 m below the circular face of the hemisphere at a constant speed of v m s^{-1}. The reaction force between P and the hemisphere is 10 N.

a Find the angle between OP and the circular face of the hemisphere.

b Find the mass of P.

c Find the value of v.

19 A particle P of mass 2 kg is attached to one end of a light inextensible string of length l m. The other end of the string is attached to a fixed point O. Initially, P is vertically below O and the string is taut. P is projected horizontally at a speed of 10 m s^{-1} such that P moves in a vertical circular arc with centre O. When OP makes an angle θ with the downward vertical, P is travelling at speed v m s^{-1} and the tension in the string is T N.

a Show that $T = 60\cos\theta + \dfrac{200}{l} - 40$

b Given that P moves in a complete circle, find the greatest value of l.

20 A particle P of mass $2m$ kg is attached to one end of a light inextensible string of length $3\,$m. The other end of the string is attached to a fixed point A. P moves at a constant angular speed of $\omega\,\text{rad}\,\text{s}^{-1}$ in a horizontal circle and AP makes an angle θ with the downwards vertical.

   a Show that $\cos\theta = \dfrac{g}{3\omega^2}$.

   b Find the range of possible values for ω.

Mathematics in life and work

New go-karts have been purchased for the theme park. You have found that the coefficient of friction between the tyres and the track is now 0.8 when the surface is dry and 0.5 when the surface is wet.

1 What is the maximum speed at which the go-karts can travel so they do not slide when going around the bend of radius 25 m?

2 By how much would this speed decrease when the track is wet?

4 HOOKE'S LAW

LEARNING OBJECTIVES

You will learn how to:

> use Hooke's law as a model relating the force in an elastic string or spring to the extension or compression

> understand the term modulus of elasticity

> use the formula for the elastic potential energy stored in a string or spring

> solve problems involving forces due to elastic strings or springs, including those where considerations of work and energy are needed.

LANGUAGE OF MATHEMATICS

Key words and phrases you will meet in this chapter:

> angular speed, compression, conical pendulum, elasticity, elastic potential energy, energy, extension, force, gravitational potential energy, kinetic energy, limiting equilibrium, mass, modulus of elasticity, tension, work done

PREREQUISITE KNOWLEDGE

You should already know how to:

> apply Newton's second law of motion

> use the equations for constant acceleration

> resolve forces on a plane and slope

> work with rough as well as smooth planes

> resolve forces in questions involving pulleys and light inextensible strings

> calculate gravitational potential energy and kinetic energy

> calculate work done against resistance

> use gravitational potential energy, kinetic energy and work done in energy calculations

> complete calculations involving conical pendulums.

You should be able to complete the following questions correctly:

1 A triangular wedge has two rough plane faces. One is inclined at 30° to the horizontal and the other at 40° to the horizontal. Two particles, A and B, of mass 4m and m, respectively, are attached to the ends of a light inextensible string. Particle A lies on the plane inclined at 40° and particle B on

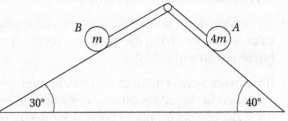

the plane inclined at 30°. The string passes over a smooth light pulley fixed on the top of the wedge. The face on which particle A moves has $\mu = 0.1$ and the face on which particle B moves has $\mu = 0.15$. The strings lie in the same vertical plane as the lines of greatest slope on each of the faces. The system is released from rest and the particles accelerate with an acceleration of $a\ \text{m s}^{-2}$. Find:

a the acceleration of the particles

b the tension in the string in terms of m

c the resultant force acting on the pulley in terms of m.

2 Two small rings, X and Y, with mass m and $2m$, respectively, are threaded on a rough horizontal pole. The coefficient of friction between each ring and the pole is μ. The rings are attached to the ends of a light inextensible string. A smooth ring Z, of mass $4m$, is threaded on the string and hangs in equilibrium below the pole. Z is in **limiting equilibrium** with angles XYZ and

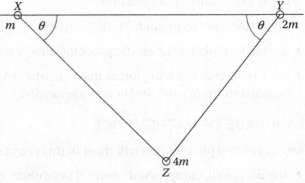

YXZ both equal to θ, where $\tan\theta = \dfrac{3}{4}$. Find:

a the tension in the string in terms of m and g

b the value of μ.

3

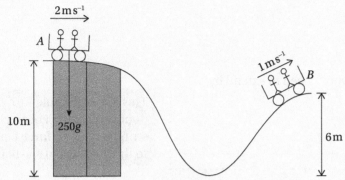

A powered theme park carriage passes point A on its route, moving downhill at $2\,\text{m}\,\text{s}^{-1}$. When the carriage has ascended 6 m to point B its speed is $1\,\text{m}\,\text{s}^{-1}$. The carriage and passengers have a combined mass of 250 kg. The total distance from A to B is 100 m. The non-gravitational resistances to motion are constant and have a total magnitude of 400 N. Find the work done by the carriage.

4.1 Definition of Hooke's law

Hooke's law states that the **tension** in an elastic string is proportional to the ratio of its **extension** to its natural, unstretched length.

$$T \sim \frac{x}{l}$$

where T is the tension in the string, x is the extension and l is the natural length.

This can also be written as:

$$T = \lambda \frac{x}{l}$$

where T is the tension in the string, x is the extension, l is the natural length and λ is the **modulus of elasticity**, which is a property of the string and will not change unless the string is damaged. The larger λ is, the more difficult the string is to stretch. By inspection, the units of λ are the same as the units of tension, that is, newtons (N).

For an elastic spring, the tension or thrust, T, can be found using the same equation.

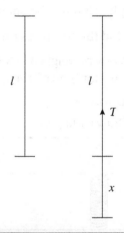

KEY INFORMATION

$T = \lambda \frac{x}{l}$

where T is the tension or thrust in the spring, x is the extension or **compression**, l is the natural length and λ is the modulus of elasticity. You are calculating a tension if the spring is stretched and a thrust if the spring is compressed.

Example 1

A string with modulus of elasticity 12 N has a natural length of 3 m.

a What is the tension in the string when it is stretched to 3.75 m?

b What tension would be required for it to double in length?

Solution

a $T = \dfrac{12 \times 0.75}{3}$

$T = 3\,\text{N}$

b To double in length the string would need to extend by 3 m so:

$T = \dfrac{12 \times 3}{3}$

$T = 12\,\text{N}$

> This is a useful rule – to double the length of an elastic string requires a **force** equal to the modulus of elasticity.

Example 2

A wobble board is created by fixing a board of weight 200 N onto two springs, S_1 and S_2, which are both secured to the floor. S_1 has a modulus of elasticity of 500 N and a natural length of 30 cm. S_2 has a modulus of elasticity of 480 N and a natural length of 35 cm. The board lies horizontally on these springs.

a Find the compression in each of the springs.

b A boy of weight 350 N steps on the wobble board. How far is he off the ground?

Solution

a Draw a diagram.

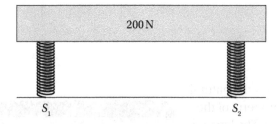

For S_1: $l = 0.3\,\text{m}$, $\lambda = 500\,\text{N}$ and $T = T_1$

For S_2: $l = 0.35\,\text{m}$, $\lambda = 480\,\text{N}$ and $T = T_2$

Resolving vertically:

$T_1 + T_2 = 200\,\text{N}$

By Hooke's law:

Let x_1 be the compression in spring 1 and x_2 be the compression in spring 2.

$T_1 = \dfrac{500x_1}{0.3}$ $\qquad\qquad$ $T_2 = \dfrac{480x_2}{0.35}$

$\dfrac{500x_1}{0.3} + \dfrac{480x_2}{0.35} = 200$

$1750x_1 + 1440x_2 = 210$

Since the board is level, the lengths of the two compressed springs must be the same. To allow this, the compressions must be different, so:

$0.3 - x_1 = 0.35 - x_2$

$x_1 = x_2 - 0.05$

Hence:

$1750(x_2 - 0.05) + 1440x_2 = 210$

$3190x_2 = 297.5$

$x_2 = 0.0933\,\text{m}$

$x_1 = 0.0433\,\text{m}$

b For S_1: $l = 0.3\,\text{m}$, $\lambda = 500\,\text{N}$ and $T = T_1$

For S_2: $l = 0.35\,\text{m}$, $\lambda = 480\,\text{N}$ and $T = T_2$

Resolving vertically:

$T_1 + T_2 = 550\,\text{N}$

By Hooke's law:

Let x_1 be the compression in spring 1 and x_2 be the compression in spring 2.

$$T_1 = \frac{500x_1}{0.3} \qquad\qquad T_2 = \frac{480x_2}{0.35}$$

$$\frac{500x_1}{0.3} + \frac{480x_2}{0.35} = 550$$

$1750x_1 + 1440x_2 = 577.5$

$1750(x_2 - 0.05) + 1440x_2 = 577.5$

$3190x_2 = 665$

$x_2 = 0.208\,\text{m}$

$x_1 = 0.158\,\text{m}$

The boy is $0.3 - 0.158 = 0.142\,\text{m} = 14.2\,\text{cm}$ above the ground.

Stop and think What conclusions can you draw about the position of the two springs?

Example 3

Party decorations are being hung to celebrate a birthday. An elastic string of natural length 25 cm is fixed to points A and B, which are 30 cm apart. A decoration of **mass** 0.6 kg is fixed to the midpoint of the string and hangs in equilibrium, at point C, 3 cm below the midpoint (M) of line AB. Calculate the modulus of elasticity of the string, stating any assumptions used.

Solution

Draw a diagram.

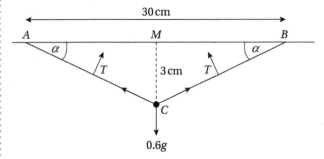

Model the decoration as a particle.

By Pythagoras' theorem:

$AC^2 = MC^2 + AM^2$

$AC = \sqrt{3^2 + 15^2}$

$AC = 15.30$

Now resolve to find the tension in the string.

Resolving vertically:

$2T \sin \alpha = 0.6g$

$2T \times \dfrac{3}{15.3} = 0.6 \times 10$

$T = 15.3$

Using Hooke's law:

$\lambda = \dfrac{1.3 \times 25}{5.3}$

The modulus of elasticity = 72.2 N

Example 4

A particle of mass 0.7 kg is attached to one end of a light elastic string on a sloping rough plane inclined at an angle of 40°. The other end of the string is fixed at point B on the slope. The natural length of the string is 1 m and its modulus of elasticity is 20.3 N. The particle is held at rest on the slope at a point 1.8 m from B. Given that the coefficient of friction is 0.3, find the initial acceleration of the particle.

Solution

Draw a diagram.

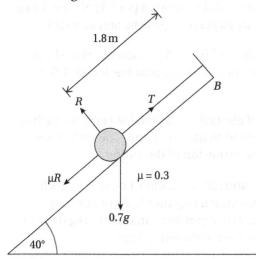

Let the tension in the string be T N.

Using Hooke's law:

$$T = \frac{20.3 \times 0.8}{1}$$

$T = 16.24$

Resolving perpendicular to the plane:

$R = 0.7g \cos 40°$

$R = 5.362$ N

Using Newton's second law:

$F = ma$

T – friction – parallel component of weight $= 0.7a$

$16.24 - 0.3 \times 5.362 - 0.7 \times 10 \sin 40° = 0.7a$

$10.13 = 0.7a$

$a = 14.5 \, \text{m s}^{-2}$

Exercise 4.1A

1. An elastic string has natural length 3 m and modulus of elasticity 40 N. Find the tension in the string when the extension is 0.2 m.

2. The tension in an elastic string is 37 N. The modulus of elasticity is 60 N. The length of the stretched string is 1.4 m. Find the natural length of the string.

3. A spring of natural length 1.2 m exerts a thrust of 33 N when compressed to a length of 0.9 m. Calculate the modulus of elasticity of the spring.

4 A spring is compressed to 0.75 of its natural length. The modulus of elasticity of the spring is 32 N. Calculate the force in the spring.

5 A spring with modulus of elasticity 120 N is fixed to the floor and supports a particle of mass 1.4 kg at the other end. Calculate the compression of the spring as a percentage of its natural length.

6 An elastic string with modulus of elasticity 30 N has one end fixed. A particle of mass 2 kg is hanging in equilibrium from the free end of the string and the string now has length 1.5 m. Find the natural length of the string.

7 A light spring of natural length 0.2 m has a modulus of elasticity of 350 N. One end is attached to a fixed point O, and a particle of mass 5.5 kg is attached to the other end. The spring is in equilibrium and hangs vertically from O. Calculate the extension of the spring.

8 Two elastic strings, S_1, with modulus of elasticity 17 N and natural length 1 m, and S_2, with modulus of elasticity 20 N and natural length 0.6 m, are joined together to form one long string which then has one end fixed to the ceiling. The other end has a mass of 0.5 kg fixed to it, which hangs in equilibrium. Calculate the length of the combined string.

9 A particle P of mass 2 kg is attached to two light elastic strings, each of natural length 0.75 m and modulus of elasticity 45 N. One string is attached to a point S on a ceiling and the other to point T 2.7 m below S. Calculate the tension in string SP and the tension in string TP.

(MM) (PS) 10 X and Y are two fixed points on a smooth horizontal table, with $XY = 1.5$ m. A particle P is attached to X by a light elastic string of natural length 0.5 m and modulus of elasticity λ, and to Y by a light elastic string of modulus of elasticity 3λ and natural length 0.3 m. Given that the particle is in equilibrium, and taking g to be 9.8, find the extensions in the two strings.

11 An object of mass 4 kg is attached to one end A of an elastic string of natural length 50 cm and modulus of elasticity 130 N. The other end of the string is attached to a point B, on a smooth inclined plane. Given that the string lies in equilibrium on the line of greatest slope and $AB = 60$ cm, calculate the tension in the string and the angle of inclination of the plane.

12 A and B are two points on a smooth horizontal floor, where $AB = 7$ m. A particle P of mass 0.6 kg is attached to point A by a light elastic spring of natural length 3 m and modulus of elasticity 16 N. It is attached to point B by a light elastic spring of natural length 2 m and modulus of elasticity 14 N. Find the extensions in the two springs when the particle rests in equilibrium.

Stop and think What would happen if the particle was pulled 1 m towards A and released? What would the motion of the particle look like?

(PS) 13 A and B are two fixed points 1.5 m apart, on a rough table with a coefficient of friction of value 0.3. A particle, P, of mass 1 kg is held in equilibrium by two light elastic strings. AP is a string with modulus of elasticity λ and natural length 0.5 m, and BP is a string of modulus of elasticity 2λ and natural length 0.75 m. Given the extension in BP is 0.15 m, calculate the value of λ, taking g to be 9.8.

14 A particle of mass 2 kg is attached to one end of a light elastic string of natural length 0.3 m and modulus of elasticity 60 N. The other end of the string is attached to a fixed point. The particle is pulled aside by a horizontal force F and it hangs in equilibrium. Given that the string then has length 0.4 m, and taking $g = 9.8\,\text{ms}^{-2}$, calculate F and the angle the string makes with the vertical.

> **Stop and think**   What would be wrong with the question if the modulus of elasticity was 30 N?

15 A particle P of mass 500 g is attached to two identical springs, S_1 and S_2, with modulus of elasticity 25 N and natural length 1 m. S_1 is attached at one end to P and at the other end to the ceiling. S_2 is attached at one end to P and at the other end to the floor. The floor and the ceiling are 2.6 m apart. Assuming that both springs are in tension and that P is in equilibrium, calculate the tensions in both springs and the height of P from the floor.

(MM) (PS) 16 A particle of mass 4 kg is attached to one end S of a spring ST of modulus of elasticity 100 N and natural length 0.5 m. End T of the spring is attached to a point on a rough surface inclined at an angle of $\sin^{-1}(0.8)$ to the horizontal. The particle rests in equilibrium, but at the point of moving down the slope, on the surface with ST along a line of greatest slope with S lower than T. A horizontal force of magnitude 40 N is applied to the particle and has caused the spring to compress to a length of 0.3 m. Taking g to be 9.8, calculate the coefficient of friction between the particle and the slope.

(PS) 17 A metal box of mass 0.5 kg is placed on a rough horizontal table. It is being pulled, horizontally, by a stretched light elastic string with natural length 20 cm and modulus of elasticity of 40 N. The coefficient of friction between the box and the table is 0.5. The acceleration of the box is 25 cm s^{-2}. Taking g to be 9.8, calculate the length of the string.

Mathematics in life and work: Group discussion

You work for a building company that has been awarded the contract to rebuild a neighbourhood playground. You have been asked to choose the spring and overall design for a rocking horse ride.

A survey of local parents has come back with the following concerns.

› Very small children, of mass less than 12 kg and height between 0.6 m and 0.8 m, need their parents to hold them and parents were finding the springs were often too short.

› Children of mass between 12 kg and 15 kg and height between 0.8 m and 1.0 m often found that although they could sit on the ride without their parents helping them, the spring didn't compress enough for them to have fun.

› Children of mass between 15 kg and 20 kg and height over 1.2 m wanted to go on the ride but the spring would compress too much and they also were unable to have fun with the ride.

1 Given this information, is it possible to find a spring with the correct height and modulus of elasticity for all the children to enjoy themselves?

2 Your project manager asks you to put together a presentation that he will use to explain to local residents how he has responded to their feedback. Answer the following questions for the residents, who are not all mathematicians and engineers.

a Is the spring realistic?

b Have you considered the minimum compression of the spring?

c How are you modelling the children (and the adults)?

> **Stop and think**   What about getting off the ride? What may happen? After studying the next section it may be worth revisiting and checking if it is possible to dismount from the seat safely, or if your choice of spring may make a fall much worse.

4.2 Elastic potential energy

From Mechanics 1, Chapter 5 you know that:

Work done = force × distance moved in the direction of the force

For an elastic string, using Hooke's law:

$$T = \lambda \frac{x}{l}$$

when the extension is x.

The work done to extend the string by a further small amount δx is given by:

Work done = $T\delta x$

Hence the total work done extending the string from $x = 0$ to $x = X$ will be given by:

Work done = $\int_0^X T \, dx$

Work done = $\int_0^X \lambda \frac{x}{l} \, dx$

Work done = $\left[\frac{\lambda x^2}{2l} \right]_0^x$

Work done = $\frac{\lambda X^2}{2l}$

According to the work–**energy** principle, the **elastic potential energy** stored in a string or spring extended by length x from its natural length l will be equal to the work done in stretching the spring or string.

Hence:

Elastic potential energy = $\frac{\lambda x^2}{2l}$

> **KEY INFORMATION**
>
> Work done stretching or compressing a spring of natural length l and modulus of elasticity λ by an amount x:
>
> Work done = $\frac{\lambda x^2}{2l}$
>
> Given that λ is in newtons and x and l are in metres, the work done will be measured in joules.

> The proof of this formula will not be required in the exam.

> **KEY INFORMATION**
>
> Elastic potential energy = $\frac{\lambda x^2}{2l}$

Example 5

A spring of natural length 1.6 m and modulus of elasticity 70 N is initially compressed to a length of 0.7 m. It is then further compressed to a length of 0.2 m. Calculate the increase in the energy stored in the spring.

Solution

Initial compression:

Elastic potential energy $= \dfrac{70 \times 0.9^2}{2 \times 1.6} = 17.72\,\text{J}$

Final compression:

Elastic potential energy $= \dfrac{70 \times 1.4^2}{2 \times 1.6} = 42.875\,\text{J}$

Hence the increase in energy stored in the spring is
$42.875 - 17.72 = 25.2\,\text{J}$ (3 s.f.)

Example 6

A particle of mass 0.6 kg is being held up to the ceiling. It is attached to the ceiling by a light elastic string of natural length 1.2 m and modulus of elasticity 7 N. The particle is released. Calculate the length of the string at its maximum extension.

Solution

Draw a diagram to visualise what will be happening to the particle. At maximum extension the speed will be $0\,\text{m s}^{-1}$.

$0\,\text{m s}^{-1}$ means instantaneous rest.

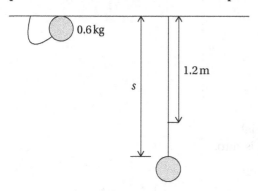

Let the length of the string at its maximum extension be s.
The extension will be $(s - 1.2)\,\text{m}$.

Elastic potential energy $= \dfrac{7 \times (s - 1.2)^2}{2 \times 1.2}$

Gravitational potential energy lost $= mgh$

$$= 0.6 \times 10 \times s$$

Due to conservation of energy and because initial **kinetic energy** and final kinetic energy are zero:

elastic potential energy = gravitational potential energy lost

$$\frac{7 \times (s - 1.2)^2}{2 \times 1.2} = 6s$$

$$7(s - 1.2)^2 = 14.4s$$

$$7s^2 - 16.8s + 10.08 = 14.4s$$

$$7s^2 - 31.2s + 10.08 = 0$$

$$s = \frac{31.2 \pm \sqrt{(31.2^2 - 4 \times 7 \times 10.08)}}{2 \times 7}$$

$$s = 4.11 \text{ or } 0.35$$

Therefore the length of the string when the particle reaches its lowest point is 4.11 m.

Stop and think   Why can't s be $0.35\,m$?

Example 7

A particle P, of mass 1.7 kg, is attached to one end of a light elastic string of natural length 1.2 m and modulus of elasticity 30 N. The other end of the string is fixed to a point A on a rough, horizontal table with $\mu = 0.2$. P is released from rest at a point 3 m from A. Calculate the speed of P when the string becomes slack.

Solution

Draw a diagram.

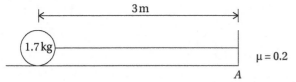

When the string is slack the particle has no elastic potential energy. This happens at the point at which the extension is zero.

The extension is zero when the string is back at its natural length of 1.2 m. So the particle will have moved 1.8 m.

Let the speed of P when $AP = 1.2$ m be v.

The elastic potential energy when the string is held at its initial position is:

$$\frac{30 \times 1.8^2}{2 \times 1.2} = 40.5 \text{ J}$$

Friction $= \mu R = 0.2 \times 1.7g = 3.4 \text{ N}$

So work done against friction = force × distance

$$= 3.4 \times 1.8 = 6.12 \, J$$

Initial kinetic energy is zero.

By the work–energy principle:

final kinetic energy	+	work done against friction	=	loss of elastic potential energy

$$\frac{1}{2} mv^2 + 6.12 = 40.5$$

$$\frac{1}{2} \times 1.7 \times v^2 = 40.5 - 6.12$$

$$v^2 = \frac{40.5 - 6.12}{0.85}$$

$$v = 6.36 \, \text{m s}^{-1}$$

> Always check for initial KE not being zero.

Example 8

A child's toy works by loading a car of mass 300 g into a spring-loaded launch box, which projects the car along a rough table where $\mu = 0.1$.

The elastic spring has natural length 10 cm and modulus of elasticity 20 N. It is compressed to 4 cm on loading. Find the speed of the car when the spring is 8 cm in length.

Solution

The elastic potential energy of the spring when fully compressed is given by:

$$\frac{20 \times 0.06^2}{2 \times 0.1} = 0.36 \, J$$

The elastic potential energy of the spring when it has reached 8 cm is given by:

$$\frac{20 \times 0.02^2}{2 \times 0.1} = 0.04 \, J$$

Work done against friction = $0.1 \times 0.3g \times 0.04 = 0.012 \, J$

Initial kinetic energy = 0

By the work–energy principle:

final kinetic energy	+	work done against friction	=	loss of elastic potential energy

$$\frac{1}{2} mv^2 + 0.012 = 0.36 - 0.04$$

$$\frac{1}{2} \times 0.3 \times v^2 = 0.32 - 0.012$$

$$v^2 = \frac{0.308}{0.15}$$

$$v = 1.43 \, \text{m s}^{-1}$$

Example 9

A light elastic string of natural length 2 m and modulus of elasticity 37.3 N has one end attached to a fixed point O. A particle of mass 3 kg is attached to the other end. The particle is held at a point which is 5 m vertically below O and released from rest. Find the distance from O at which the particle first comes to rest.

Note that the question is not asking about where the particle finally comes to rest, but where it *first* comes to rest – that is, when $v = 0$.

Solution

You first need to find the speed of the particle at the point where the spring becomes slack so that you know how much kinetic energy it has at that point.

Conservation of energy from when the particle is released to the moment the string becomes slack gives:

gain in kinetic energy $+$ gain in potential energy $=$ loss in elastic energy

$$\frac{1}{2} \times 3 \times v^2 + 3 \times 10 \times 1 = \frac{37.3 \times 3^2}{2 \times 2}$$

$$1.5v^2 = 83.925 - 30$$

$$v^2 = 35.95$$

$$v = 6.00 \, \text{m s}^{-1}$$

You now know that the particle is travelling at $6.00 \, \text{m s}^{-1}$. The particle comes to rest when $v = 0$.

By conservation of energy:

kinetic energy loss = potential energy gain

$$\frac{1}{2} mv^2 = mgh$$

$$35.95 = 2 \times 10 \times h$$

$$h = 1.80 \, \text{m}$$

So the particle is $5 - 2 - 1.80 = 1.2$ m from O.

Stop and think   What happens after the particle first comes to instantaneous rest? What would the motion look like?

Mathematics in life and work: Group discussion

You are designing a range of home gym equipment. You are asked to make a chest expander from a number of springs of natural length 28 cm and modulus of elasticity 210 N, attached to two handles.

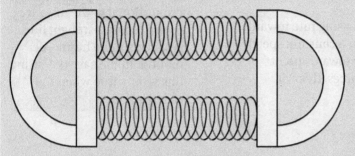

1 How many springs would you need so that a force of roughly 650 N would extend it by 40 cm?

Another piece of equipment is made up of five springs of natural length 0.9 m and modulus of elasticity 170 N. One end is held to the ground by the user's foot and the other is attached to a rod of mass 1 kg that the user lifts up.

2 What force would it require to lift it to your head height? How much energy would you use by lifting this rod to a height of 2 m for 10 repetitions? If you were to let it slip from your hand, how long would it take for the springs to start to be compressed?

Example 10

A particle of mass 0.5 kg rests on a smooth horizontal table attached to one end of an elastic string of natural length 0.8 m and modulus of elasticity 20 N. The other end of the string is fixed to a point O on the table. The particle is pulled to a point 1 m from O and released.

a Find the speed of the particle when the string returns to its natural length.

b Calculate the time between the string becoming slack and it becoming taut again.

c Describe the motion of the particle.

Solution

a EPE when extended $= \dfrac{20 \times 0.2^2}{2 \times 0.8} = 0.5$J

$$\frac{1}{2} \times 0.5v^2 = 0.5$$
$$v^2 = 2$$
$$v = 1.41\,\mathrm{m\,s^{-1}}$$

b The string is slack for $2 \times 0.8 = 1.6\,\text{m}$

$$\text{Time} = \frac{\text{distance}}{\text{speed}}$$

$$\text{Time} = \frac{1.6}{\sqrt{2}} = 1.13\,\text{s}$$

c The particle will accelerate at a decreasing rate towards the point O for 0.2 m and then travel at a constant speed for 1.6 m, after which it will accelerate towards point O again. As there are no resistance forces, this will continue.

> This oscillating motion, where the acceleration is towards a fixed point O and is proportional to the distance from O along a straight line, is called simple harmonic motion. In the case of a string, this is only true when the string is taut.

Example 11

A particle P, of mass 5 kg, is attached to one end of a light elastic string of natural length 0.7 m and modulus of elasticity 30 N. The other end of the string is fixed to a point O on a rough plane, which is inclined at an angle of 40° to the horizontal. The particle is held at O and then released from rest. It stops after moving 1.5 m down the plane. Taking g to be $9.8\,\text{m}\,\text{s}^{-2}$, find the coefficient of friction between the particle and the plane.

Solution

Draw a diagram.

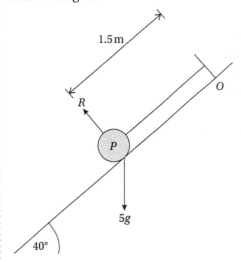

Resolving perpendicular to the plane:

$$R = 5g\cos 40$$

The work–energy principle gives:

elastic potential energy gained $=$ potential energy loss $-$ work done against friction

You need to calculate the vertical drop through which the particle has moved. This will be 1.5 sin 40°.

$$\frac{30 \times 0.8^2}{2 \times 0.7} = 5 \times 9.8 \times 1.5 \sin 40° - \mu \times 5g \cos 40°$$

$$\frac{96}{7} = 73.5 \sin 40° - 49\mu \cos 40°$$

$$\mu = \frac{73.5 \sin 40° - \frac{96}{7}}{49 \cos 40°}$$

$$\mu = 0.893 \text{ (3 s.f.)}$$

In Chapter 3, you encountered objects moving in a horizontal circle, where the string traces the outline of a cone, giving you a conical pendulum. You modelled these using light inextensible strings. Hooke's law allows you to study conical pendulums when the strings are elastic.

Example 12

A small mass B has a mass of 1.5 kg. It is attached to one end of a light elastic string of natural length 0.6 m and modulus of elasticity 1200 N. The other end of the string is attached to a fixed point A. The mass is moving with constant speed in a horizontal circle with its centre vertically below A. This creates a **conical pendulum**, since the string describes the surface of a cone as the mass moves in a circle. The string is inclined at an angle of θ to the downward vertical and $\sin \theta = \frac{3}{5}$. Calculate the stretched length of the string and hence find the **angular speed** of the mass.

Recall that the acceleration of a particle moving in a circle is calculated using the formula $a = r\omega^2$ and thus the resultant force is calculated using the formula $F = mr\omega^2$.

Solution

Draw a diagram.

$\sin \theta = \frac{3}{5}$

$\cos \theta = \frac{4}{5}$

Resolving vertically:

$T \cos \theta = 1.5g$

From Hooke's law:

$T = \frac{x}{0.6} \times 1200$

So: $\frac{x}{0.6} \times 1200 \cos \theta = 1.5g$

$\qquad 1600x = 15$

$\qquad x = 0.009\,375$

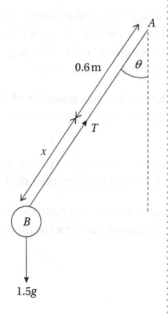

Therefore the length of the stretched string is 0.609 m (3 s.f.).

By Newton's second law:

$F = ma$

$T \sin \theta = 1.5 r \omega^2$

From above:

$T = \dfrac{0.009\,375}{0.6} \times 1200 = 18.75$

$r = 0.609 \sin \theta = 0.365$

so

$18.75 \times 0.6 = 1.5 \times 0.365 \times \omega^2$

$\omega^2 = 20.548$

$\omega = 4.53 \, \text{rad s}^{-1}$

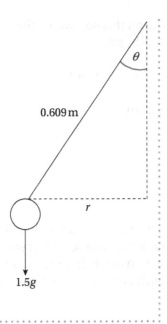

Exercise 4.2A

1. Calculate the work done when an elastic spring of length 0.5 m and modulus of elasticity 8 N is doubled in length.

2. Calculate the elastic potential energy stored in a spring of natural length 0.7 m and modulus of elasticity 7 N is compressed to 0.2 m.

(PS) 3. A light elastic spring has natural length 1 m and modulus of elasticity 15 N. One end is attached to the ceiling and the other to a particle of mass 2 kg. Calculate the energy stored in the spring.

4. A light elastic spring of natural length 2 m and modulus of elasticity 50 N rests on a smooth horizontal floor with one end fixed. A particle of mass 1 kg is attached to the free end of the spring, which is compressed to a length of 0.5 m. Taking g to be 9.8, calculate the speed of the particle when the spring is at its natural length of 2 m.

5. A light elastic string of natural length 1 m has modulus of elasticity 2000 N. The work done stretching the string is 30 J. Find the extension and hence the stretched length of the string.

(PS) 6. A spring of modulus of elasticity 48 N is stretched to four times its natural length. Given that the work done achieving this is 60 J, find the natural length of the spring.

7. A spring of natural length 2 m and modulus of elasticity 100 N is initially compressed to a length of 0.6 m. It is then further compressed to a length of 0.1 m. Calculate the increase in the energy stored in the spring.

8 A particle P, of mass 2.5 kg, is attached to one end of a light elastic string of natural length 1 m and modulus of elasticity 50 N. The other end of the string is fixed to a point A on a rough horizontal table with $\mu = 0.3$. P is released from rest at a point 5 m from A. Calculate the speed of P when the string becomes slack.

9 A small mass B has a mass of 4 kg. It is attached to one end of a light elastic string of length 0.75 m and modulus of elasticity 2000 N. The other end of the string is attached to a fixed point A. The mass is moving with constant speed in a horizontal circle with centre vertically below A, creating a conical pendulum. The string is inclined at an angle 40° to the vertical. Calculate the stretched length of the string and hence find the angular speed of the mass.

10 A particle P has a mass of 2 kg and is attached to one end of a light elastic string of natural length 1.7 m and modulus of elasticity 5 N. The other end is attached to a fixed point on a smooth horizontal table 3 m from where P is being held, before it is released. Calculate the speed of P when the string becomes slack.

(C) (MM) (PS) 11 A light elastic string of natural length 1.5 m and modulus of elasticity 20 N is attached at one end to a ceiling at point X. The other end of the string is attached to a particle of mass 1.3 kg. The particle is held at a distance of 1 m vertically below X and released. Find the length of the string when the particle reaches its lowest point and the speed of the particle when it passes through its equilibrium position.

12 A rope is stretched across a river of width 340 m. Assuming that the rope is elastic with modulus of elasticity of 1000 N and natural length 300 m, calculate the work done in stretching the rope.

13 A particle of mass 0.7 kg is attached to one end of a light elastic string of natural length 0.5 m. The other end of the string is attached to a fixed point A. The particle is released from rest at A and comes to instantaneous rest 2 m below A. Find the modulus of elasticity of the string.

(MM) (PS) 14 A jack-in-the-box toy is made of a spring of natural length 30 cm and modulus of elasticity 80 N. The doll has a mass of 0.6 kg. When the lid is closed the spring is compressed to a length of 10 cm. Calculate the maximum distance the doll will rise when the lid is released. What assumptions were necessary?

(C) 15 A particle of mass m is attached to one end of a light elastic string with natural length l and modulus of elasticity λ. The other end of the string is attached to a fixed point A on a smooth plane inclined at θ, such that $\sin \theta = \dfrac{5}{13}$. The particle is held at rest along the line of greatest slope. When released it first comes to instantaneous rest after moving x m. Find x in terms of m, l and λ.

(MM) 16 A toy car of mass 200 g is attached to an elastic string of natural length 2 m and modulus of elasticity 8 N. The other end of the string is attached to the base of a wall 4 m away. Assuming that the floor is smooth and the resistance force is 1 N, find the speed at which the car crashes into the wall.

(PS) **17** A man of mass 86 kg has decided to complete a bungee jump from a platform in a car park. For safety reasons, the organisers want him to fall no further than 70 m. Given that the bungee rope has a modulus of elasticity of 5000 N, what is the maximum value for the rope's natural length?

(MM) **18** Li makes a catapult by attaching a piece of elastic of natural length 0.2 m and modulus of elasticity 50 N to a Y-shaped stick. The points of attachment are 0.15 m apart. He takes a stone of mass 0.1 kg, pulls the elastic back until it is 30 cm behind the catapult and releases it horizontally. Calculate the speed of the stone on release.

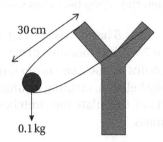

30 cm

0.1 kg

(MM) **19** A model aeroplane of mass 0.4 kg is attached to the ceiling of a bedroom with a light elastic
(PS) string of natural length 40 cm and modulus of elasticity 15 N. The aeroplane is accidentally pulled down until it is 1 m from the ceiling and released. Calculate the speed at which the aeroplane is travelling when the string becomes slack and find out if the aeroplane hits the ceiling.

(PS) **20** A particle of mass 5 kg is attached to one end of a light elastic string of natural length 1 m and modulus of elasticity 9400 N. One end of the string is attached to a point X, 2 m above the floor. The particle is raised to 1 m above X and released. Find its velocity at the point 1 m below X. Calculate the length of the string when the particle first comes to instantaneous rest.

(MM) **21** A safety feature on a theme park ride ensures that, should a rollercoaster carriage not have left the next stage, the following one is slowed to a stop by means of an elastic rope of natural length 5 m and modulus of elasticity 2000 N, fixed at point O, which attaches to the following carriage. The carriage has a mass of 300 kg and will be travelling at $10 \, \text{m s}^{-1}$ if the safety feature needs to deploy. Once the safety feature has been deployed and the carriage brought to rest, it will be fixed in place and winched to the next phase when it is safe to do so.

a Find the kinetic energy of the carriage as it passes point O.

b Find how far the carriage travels from O before it is brought to rest.

c What assumptions have you made?

SUMMARY OF KEY POINTS

❯ Hooke's law states that the tension or thrust in an elastic string or spring is proportional to the ratio of the extension to the natural length:

$$T = \lambda \frac{x}{l}$$

where λ is the modulus of elasticity, x is the extension and l is the natural length of the string or spring.

❯ Hooke's law calculates the tension if the string or spring is stretched and the thrust if the spring is compressed.

❯ The work done in stretching or compressing an elastic string or spring can be calculated using integration and gives the formula:

work done $= \dfrac{\lambda x^2}{2l}$

where λ is the modulus of elasticity, x is the extension or compression and l is the natural length of the string or spring.

❯ The elastic potential energy stored in a string or spring when it is stretched or compressed is given by:

$\text{EPE} = \dfrac{\lambda x^2}{2l}$

where λ is the modulus of elasticity, x is the extension or compression and l is the natural length of the string or spring.

❯ When considering the changes of energy within a system involving springs and elastic strings, you always need to consider gravitational potential energy, kinetic energy and elastic potential energy. Using the principle of conservation of energy to create equations is a good starting point for most of your calculations involving Hooke's law.

❯ Maximum extension or compression will occur when $v = 0\,\text{m}\,\text{s}^{-1}$ (instantaneous rest).

❯ At maximum speed the acceleration will be $0\,\text{m}\,\text{s}^{-2}$.

EXAM-STYLE QUESTIONS

1 The points A and B are on a smooth horizontal table at a distance 50 cm apart. A particle P, of mass 2 kg, lies on the table on the line AB, between A and B. P is attached to the midpoint of a light elastic string of natural length 0.4 m and modulus of elasticity λ N. One end of the string is attached to A and the other end is attached to B. P is in equilibrium at the point O on AB, with a tension in the string of 300 N.

   a Find the value of λ.

The particle is now held at the point C where C is 10 cm away from O on the line perpendicular to AB. The particle is released.

   b Find the magnitude of the initial acceleration of the particle.

   c Find the speed of the particle as it passes through O.

2 The points A and B are on a horizontal line a distance $10\,\text{m}$ apart. A particle P, of mass m kg, is attached to the midpoint of a light elastic string of natural length $6\,\text{m}$ and modulus of elasticity $60\,\text{N}$. One end of the string is attached to A and the other end is attached to B. P is in equilibrium at a point $4\,\text{m}$ below AB.

a Calculate the value of m.

b Calculate the elastic energy of the string when it is in the equilibrium position.

③ 3 A particle P, of mass m kg, is attached to one end of a light elastic string of natural length $l\,\text{m}$ and modulus of elasticity $2mg\,\text{N}$. The other end of the string is attached to a fixed point A. The particle P is released from rest at A and falls vertically. When the particle has fallen a distance of $(l + x)\,\text{m}$, where $x > 0$, the speed of the particle is $v\,\text{m\,s}^{-1}$.

a Show that $v^2 = 2g(l + x) - \dfrac{2gx^2}{l}$.

b Find the greatest speed attained by the particle when it falls, in terms of g and l.

After release, the particle is first at rest at point Z.

c At Z, find the value of x in terms of l.

4 A particle P, of mass $6\,\text{kg}$, is attached to one end of a light elastic string of natural length $0.3\,\text{m}$. The other end of the string is attached to a fixed point O. A horizontal force of magnitude $45\,\text{N}$ is applied to P and the string is stretched beyond its natural length. Given that P is in equilibrium and that the energy stored in the string is $8\,\text{J}$, calculate the extended length of the string.

③ 5 A particle of mass $4m$ is attached to the ends of two light elastic strings of natural length $2l$ and modulus of elasticity λmg. The ends of the strings are attached to fixed points on the same horizontal level $6l$ apart. The particle hangs in equilibrium at a distance $\dfrac{\sqrt{13}}{2}l$ below the midpoint, C, between the two fixed points.

a Show that $\lambda = \dfrac{56}{3\sqrt{13}}$.

The particle is now pulled vertically downwards to a distance $12l$ below C and released from rest.

b Find v^2 for the particle, in terms of g and l, as it reaches a point level with the two fixed points.

6 One end of a light elastic string of natural length l and modulus of elasticity $5mg$ is fixed at a point X on a smooth plane inclined at $30°$ to the horizontal. A particle of mass $2m$ is attached to the other end of the string. Initially the particle is held at rest, with the string just taut, and lies below X on the plane along the line of greatest slope. The particle is released and comes to instantaneous rest at a point Y on the plane.

a Calculate the length XY in terms of l.

b Calculate the greatest speed attained by the particle as it moves from its starting position to Y, in terms of l and g.

7 One end of a light elastic string of natural length 2 m and modulus of elasticity 19.6 N is attached to a fixed point O on a ceiling. A particle of mass 0.9 kg is attached to the other end of the string. The particle is held at rest at O and then released. The particle comes to instantaneous rest for the first time at point A.

  a Find the distance OA.

  b Find the magnitude of the acceleration at A.

PS **8** A particle P, of mass $2m$, is attached to one end of a light elastic string of natural length l and modulus of elasticity $3mg$. The particle lies on a plane inclined at 60° to the horizontal and the other end of the string is attached to a fixed point O on the plane, above the particle.

The particle is in equilibrium at a point A on the plane and the extension in the string is $\frac{1}{2}l$. The particle is now projected down the plane with speed U. It comes to instantaneous rest after moving a distance of l.

  a Find U in terms of l and g.

  b Find, in terms of l and g, the speed of P when the string first becomes slack.

C **9** A light elastic string of natural length $5l$ and modulus of elasticity λ has its ends attached to two points X and Y with a particle of mass $2m$ attached to its midpoint. $XY = 4l$, XY is horizontal and the particle rests in equilibrium a distance of $3l$ below XY.

Show that $\lambda = \dfrac{65mg}{78 - 15\sqrt{13}}$.

PS **10** A light elastic spring of natural length l and modulus of elasticity λ has a toy car of mass m attached to one end. The other end of the spring is fixed to a launcher on a smooth table. The car is pushed into the launcher and the spring is compressed to half of its length and released. At the point where the spring has reached its natural length, the car is travelling at a speed of $\sqrt{2gl}$.

  a Find λ.

The launcher is rotated to a vertical position and again the car is fixed to one end. The other end is placed on the floor. The spring is compressed to a length $\frac{1}{4}l$ and released. The particle moves vertically upwards and will be travelling at speed v when the spring has reached its natural length.

  b Find v.

MM **11** A particle of mass 0.9 kg is attached to one end of a light elastic spring of natural length 1.3 m and modulus of elasticity λ. The other end of the string is attached to a fixed point A on the ceiling. The particle hangs in equilibrium 1.75 m vertically below A.

  a Find the value of λ.

The particle is raised and the spring is compressed to a point vertically below A such that the distance from A to the particle is now 0.9 m. The particle is released and first comes to instantaneous rest at the point B.

  b Find the distance AB.

MM **12** One end of a light elastic string of natural length 0.5 m and modulus of elasticity 200 N is attached to a particle of mass 20 kg. The other end is attached to a point O on the ceiling.

The particle is pulled down to a distance 1.5 m below O and released.

a Calculate the elastic potential energy stored in the string when the particle is in the position 1.5 m below O and state its extension from the equilibrium point.

b Find the distance below O the first time the particle comes to instantaneous rest.

c What assumptions are necessary?

13 A particle P, of mass 1.5 kg, is attached to one end of a light elastic spring of natural length 2 m and modulus of elasticity λ N. The other end is attached to a fixed point O. P is initially hanging freely in equilibrium 2.5 m vertically below O. P is raised vertically upwards to the point where $OP = 1.3$ m. P is then released from rest and moves vertically downwards until it comes to instantaneous rest again at the point A.

a Find the value of λ.

b Find the distance OA.

14

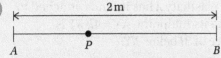

The diagram shows a particle P, of mass 2 kg, attached to one end of a light spring of natural length 0.6 m and modulus of elasticity 3 N. The other end of the spring is attached to a fixed point A. The particle P is also attached to one end of a second spring of natural length 0.9 m and modulus of elasticity 2 N. The other end of the second spring is attached to a fixed point B. A and B are fixed to a smooth horizontal table with $AB = 2$ m and APB is a horizontal straight line.

a Find the distance AP, when P is in equilibrium.

The particle P is pulled towards A and held at the point where $AP = 0.4$ m. P is then released from rest and is next at instantaneous rest at the point C.

b Find the distance AC.

PS **15** A spring of natural length 0.3 m and modulus of elasticity 50 N is fixed to a point O on the ceiling. The other end is attached to a mass of 2.4 kg.

a Find the elastic potential energy in the spring when it is at equilibrium.

The mass splits in half, with half remaining on the spring and the other half falling to the ground.

b Find the distance of the mass remaining on the spring from O when it first comes to instantaneous rest.

MM 16 A particle of mass 2 kg is attached to one end of a light elastic string of natural length 0.9 m and modulus of elasticity 36 N. The other end is fixed to a point O on a rough table. The particle is placed on the table at point X, 4 m from point O, and is then released. It travels to point Y, 2 m from O, where it comes to instantaneous rest.

a Find the frictional force acting on the particle as it moves from X to Y.

b Find the total distance travelled by the particle after it is released from X assuming that, once it has travelled more than 6 m, the next time it comes to rest is its final stop.

C 17 A small sphere S, of mass 0.3 kg, is attached to one end of a light elastic string of natural length 0.8 m and modulus of elasticity 30 N. The other end of the string is attached to a fixed point A on the ceiling. S hangs freely in equilibrium vertically below A.

a Find the extension of the string.

S is pulled vertically downwards and held in position at point B. S is released from rest and hits the ceiling at a speed of 0.1 m s^{-1}.

b Find the distance AB.

18 A light elastic string AB of natural length 1 m and modulus of elasticity λ N is attached at A to a ceiling. End B of the string is attached to a particle P of mass 0.5 kg. P is held 3 m vertically below A and released from rest. Given that P hits the ceiling, find the range of possible values for λ.

MM 19 A particle of mass m is attached to the end of a light elastic string of natural length L and modulus of elasticity λmg N. The particle moves in a horizontal circle and the string makes an angle of 60° to the vertical when extended to a length of $1.3L$.

a Find the tension in the string.

b Find λ.

c Find the angular speed of the string.

C 20 A particle P, of mass m kg, is attached to one end of a light elastic string with natural length l and modulus of elasticity $3mg$ N. The other end of the string is attached to a fixed point A on the ceiling. P is released from rest at A and falls vertically. When $AP = l + x$, where $x > 0$, the speed of P is v m s^{-1}. After release, P next comes to instantaneous rest at B.

a Find, in terms of l and x, the maximum speed of P as it falls.

b Find, in terms of l and x, the distance AB.

Mathematics in life and work

A bungee jumper of mass 88 kg is using a bungee cord of natural length 36 m and modulus of elasticity 3645 N. One end of the cord is attached to a platform that is y metres from the ground. The other is attached by means of a harness to the jumper. Model the jumper as a particle that falls vertically with zero air resistance.

1 Find the maximum speed achieved by the jumper during the descent.

2 State the minimum value of y, giving your answer to the nearest metre.

3 Comment on your findings in parts **a** and **b** and any other modelling decisions you made.

5 LINEAR MOTION UNDER A VARIABLE FORCE

Mathematics in life and work

In this chapter, you will learn how to model motion when the force varies with time, velocity or displacement. This technique is important in careers in adventure sports and automotive design – for example:

> If you were designing a new car, you could use your knowledge of air resistance to design a car that minimises air resistance and therefore reduces the amount of fuel used.

> If you worked for an adventure sports company that takes groups of people skydiving, you would need to understand how to analyse air resistance to ensure that the parachute will slow the person down enough for a safe landing.

> If you were a scuba diver wanting to minimise the amount of air consumed on an underwater swim to a wreck, you would explore ways of configuring your kit to reduce the drag acting on you.

LEARNING OBJECTIVES

You will learn how to:

> solve problems that can be modelled as the linear motion of a particle under the action of a variable force, by setting up and solving an appropriate differential equation

> use $v\dfrac{\mathrm{d}v}{\mathrm{d}x}$ for acceleration, where appropriate.

LANGUAGE OF MATHEMATICS

Key words and phrases you will meet in this chapter:

> instantaneous velocity, terminal velocity

PREREQUISITE KNOWLEDGE

You should already know how to:

> apply Newton's laws of motion to the linear motion of a particle of constant mass moving under the action of constant forces

> use differentiation and integration with respect to time to solve simple problems concerning displacement, velocity and acceleration

> convert an improper function into a proper function

> integrate trigonometric functions

> integrate functions of the forms e^{ax+b}, $\dfrac{1}{ax+b}$, $\dfrac{1}{x^2+a^2}$ and $\dfrac{kf'(x)}{f(x)}$

> integrate using partial fractions

> find, by integration, a general form of solution for a first order differential equation in which the variables are separable.

You should be able to complete the following questions correctly:

1 Express $\dfrac{x}{x+1}$ as a proper fraction.

2 Find the general solution of $\dfrac{dy}{dx} = xy$.

3 Find the general solution of $\dfrac{dy}{dx} = \dfrac{\sin x}{y}$.

4 Find the general solution of $\dfrac{dy}{dx} = \dfrac{5}{e^x}$.

5 Find the general solution of $\dfrac{dy}{dx} = 4x^2 - 25$.

6 Find the general solution of $x\dfrac{dy}{dx} = 4x^2 - 25$.

7 Find the particular solution of $x\dfrac{dy}{dx} = y^2$; $y = 4$ when $x = 1$.

8 Find the particular solution of $x\dfrac{dy}{dx} = e^y$; $y = 0$ when $x = 4$.

5.1 Motion of a particle when forces acting upon it depend on time or position

In Mechanics 1, Chapter 2, you learnt that acceleration, velocity and displacement were connected by the following relationships:

> $a = \dfrac{dv}{dt} = \dfrac{d^2x}{dt^2}$ ⚬— To find acceleration you need to differentiate velocity with respect to time.

> $v = \dfrac{dx}{dt}$ ⚬— To find velocity you need to differentiate displacement with respect to time.

> $v = \int a\, dt$ ⚬— To find velocity you need to integrate acceleration with respect to time.

> $x = \int v\, dt$ ⚬— To find displacement you need to integrate velocity with respect to time.

These relationships can be used to solve problems when the forces acting upon a particle depend on the time.

Example 1

A small object of mass 2 kg is moving in a straight line with a velocity of $v\,\mathrm{m\,s^{-1}}$ given by $v = e^{2t} + 4t + 1$.

a Find the force acting on the particle at a time t.

b Find the force acting on the particle after 3 seconds.

Solution

a You know that $a = \dfrac{\mathrm{d}v}{\mathrm{d}t}$ so to find a you need to differentiate v with respect to t.

$$v = e^{2t} + 4t + 1$$

$$a = \frac{\mathrm{d}v}{\mathrm{d}t} = 2e^{2t} + 4$$

By Newton's second law, $F = ma$.

So the force acting on the small object, $F = 2\left(2e^{2t} + 4\right)$

$$F = 4e^{2t} + 8$$

b To find the force acting on the particle after 3 seconds, substitute $t = 3$ into the above formula.

$$F = 4e^{2t} + 8$$
$$= 4e^{2 \times 3} + 8$$
$$= 1622$$

The force acting on the particle after 3 seconds is 1622 N to 3 significant figures.

If the acceleration of the object varies with either velocity or displacement and not time, you need a different approach.

You know that $a = \dfrac{\mathrm{d}v}{\mathrm{d}t}$.

Using the chain rule:

$$a = \frac{\mathrm{d}v}{\mathrm{d}t} = \frac{\mathrm{d}v}{\mathrm{d}x} \times \frac{\mathrm{d}x}{\mathrm{d}t}$$

Since $\dfrac{\mathrm{d}x}{\mathrm{d}t} = v$:

$$a = v\frac{\mathrm{d}v}{\mathrm{d}x}$$

KEY INFORMATION

If you have an expression for acceleration in terms of x, you can replace a by $v\dfrac{\mathrm{d}v}{\mathrm{d}x}$ to form a differential equation, which can be solved to find the velocity in terms of x.

Example 2

A small object of mass 2 kg is released from point A in a horizontal line with an initial velocity of $5\,\text{m s}^{-1}$. The resultant force acting on the object is given by $F = 10x$, where x is the distance from A, and is acting towards A. The object is instantaneously at rest at two different points. Find the distance between these two points.

Instantaneous velocity is the velocity of an object in motion at a specific point in time.

Solution

By Newton's second law, the acceleration of the object, a, is given by:

$2a = -10x$

$a = -5x\,\text{m s}^{-2}$

The acceleration is negative because the resultant force is directed towards the starting point.

Since $a = v\dfrac{dv}{dx}$:

$v\dfrac{dv}{dx} = -5x$

Separate variables:

$\int v\,dv = \int -5x\,dx$

$\dfrac{1}{2}v^2 = -\dfrac{5}{2}x^2 + c$

Don't forget the constant of integration.

When $x = 0$, $v = 5\,\text{m s}^{-1}$ so:

$\dfrac{1}{2} \times 5^2 = -\dfrac{5}{2} \times 0^2 + c$

$c = \dfrac{25}{2}$

$\dfrac{1}{2}v^2 = -\dfrac{5}{2}x^2 + \dfrac{25}{2}$

$v^2 = 25 - 5x^2$

$v = \sqrt{25 - 5x^2}$

You want the points where the object is at rest. This is when $v = 0\,\text{m s}^{-1}$.

When $v = 0$:

$25 - 5x^2 = 0$

$x^2 = 5$

$x = \pm\sqrt{5}\,\text{m}$

The distance between the two points when the object is instantaneously at rest is $2\sqrt{5}\,\text{m}$ or $4.47\,\text{m}$.

Stop and think

If the acceleration of an object is given in terms of the displacement from a given point, you can differentiate to get an expression for the velocity in terms of the displacement. How could you use this to help you to get an expression for the displacement in terms of time?

Exercise 5.1A

 1 A small object of mass 500 g is moving in a straight line from the point O with a speed of $14\,\text{m}\,\text{s}^{-1}$. The acceleration of the object is given by $a = t^2 - 2t$.

   a Find the velocity of the object in terms of t.

   b Find the displacement of the object from the point O in terms of t.

2 The velocity of a toy car at time t is given by $v = \dfrac{5}{5t+2}$, $t \geqslant 0$. It is released from rest from the point A. Find, in terms of t:

   a the acceleration of the toy car

   b the displacement of the toy car from A.

3 The resultant force acting on a small sphere of mass 2 kg is given by $F = 10e^{-2t}$ in the direction of movement, where t is the time from release. If the sphere is released with a velocity of $50\,\text{m}\,\text{s}^{-1}$:

   a find an expression for the velocity, v, of the small sphere in terms of t

   b show that $50 \leqslant v < 52.5$.

4 The displacement of an object from a point A is given by $x = 5\sin 2\pi t$. Find:

   a the acceleration of the object when $t = 5$

   b the greatest magnitude of the acceleration of the object.

5 A small sphere is moving in a straight line away from the point A. The acceleration of the object is given by $3x - 2\,\text{m}\,\text{s}^{-2}$, where x is the displacement of the sphere from A. The small sphere is released from A at rest and travels to B which is 3 m away from A. Find:

   a the velocity of the sphere in terms of x

   b the velocity of the sphere when it arrives at B.

6 An object of mass 2 kg is moving away from a point A with a speed of $10\,\text{m}\,\text{s}^{-1}$. The resultant force acting on the object is given by $20x$ acting towards A. What distance does the object travel before it comes to rest for the first time?

7 A sphere of mass 0.5 kg is moving along a horizontal line. It passes A with a velocity of $2\,\text{m}\,\text{s}^{-1}$ and is heading towards the point B which is 5 m away from A. The resultant force acting on the sphere is directed away from A and has a magnitude of $5e^{0.5x}$. By finding v in terms of x, find the velocity of the sphere when it reaches B.

8 An object is moving along the x axis. It starts from the origin, O, at rest and moves towards A, which is in the direction of increasing x. The acceleration of the object has magnitude given by $x^3(3 - x)\,\text{m}\,\text{s}^{-2}$ acting away from O. Find:

   a an expression for v in terms of x

   b the distance from O when the object instantaneously comes to rest.

9 An object of mass 10 kg moves along a horizontal line. The displacement of the object from its starting point is x m and its velocity at any given time is given by v m s^{-1}. A driving force of $2x$ N and a resistance force of $3x^2$ N is acting on the object. If the object leaves it starting point from rest, find:

a an expression for v in terms of x

b the distance travelled by the object before it instantaneously comes to rest.

5.2 Motion of a particle when forces acting upon it depend on velocity

In this section, you will learn how to analyse the motion of a particle moving in a straight line when the forces acting upon it depend on the velocity of the particle.

The methods used to do this are very similar to those used in the Section 5.1.

To solve problems of this type, you need to find an expression for the acceleration in terms of the velocity and use either $a = \dfrac{dv}{dt}$ or

$a = v\dfrac{dv}{dx}$ to form a differential equation. Solving this differential equation will give you either an expression for the velocity in terms of time (t) or displacement (x), depending on which formula you use.

When forming a differential equation, it is important to decide which direction the force is acting in. If it is working against the motion, the force will be negative.

> **KEY INFORMATION**
>
> › If you need to write velocity in terms of time, replace a by $\dfrac{dv}{dt}$ and solve the differential equation.
>
> › If you need to write velocity in terms of displacement, replace a by $v\dfrac{dv}{dx}$ and solve the differential equation.

Example 3

An object of mass 10 kg is being pulled along a surface with a force of 200 N. The magnitude of the resistance force is $10v$ N, where v is the velocity of the object at any given time. The object has an initial velocity of 3 ms^{-1} and moves in a straight line. Find the distance moved when the object instantaneously comes to rest.

Solution

The resultant force acting horizontally on the object is $(200 - 10v)$ N.

By Newton's second law: $F = ma$

$$200 - 10v = 10a$$

$$a = 20 - v$$

$$v\frac{dv}{dx} = 20 - v$$

> The question requires the distance moved so use $a = v\dfrac{dv}{dx}$.

$$\int \frac{v}{20-v}\,dv = \int 1\,dx$$

$$\int \left(\frac{20}{20-v}\right) - 1\,dv = \int 1\,dx$$

$$-20\ln(20-v) - v = x + c$$

When $x = 0$, $v = 3$

$$-20\ln(17) - 2 = 0 + c$$

So $c = -20\ln 17 - 2$

So $x = 20\ln 17 - \ln(20-v) + 2 - v$

$$x = \ln\left(\frac{17}{20-v}\right)^{20} + 2 - v$$

The object will instantaneously come to rest when $v = 0\,\mathrm{m\,s^{-1}}$.

$$x = \ln\left(\frac{17}{17}\right)^{20} + 3 - 0$$

$$x = 3$$

The object will instantaneously come to rest when it is $3\,\mathrm{m}$ from the starting point.

$\dfrac{v}{20-v}$ is an improper fraction because the degree of the numerator is equal to the degree of the denominator. To make it a proper fraction, you need to divide v by $20-v$, using polynomial division.

$$-v+20\overline{)\begin{array}{l} \,-1 \\ v \\ \underline{v-20} \\ 20 \end{array}}$$

So $\dfrac{v}{20-v} = -1 + \dfrac{20}{20-v}$

$$= \dfrac{20}{20-v} - 1$$

Example 4

A bowling ball of mass $350\,\mathrm{g}$ is thrown along a smooth horizontal surface with an initial speed of $20\,\mathrm{m\,s^{-1}}$. A resistance force of magnitude $0.02v^2$ acts on the ball, where v is its velocity. By forming and solving a differential equation, find the velocity of the ball after it has travelled $10\,\mathrm{m}$.

Solution

By Newton's second law: $ma = F$:

$$0.35a = -0.02v^2$$

$$0.35v\frac{dv}{dx} = -0.02v^2$$

$$35v\frac{dv}{dx} = -2v^2$$

$$35\int \frac{v}{v^2}\,dv = \int -2\,dx$$

$$35\int \frac{1}{v}\,dv = \int -2\,dx$$

$$35\ln v = -2x + c$$

You are given the displacement so use $a = v\dfrac{dv}{dx}$.

When $x = 0$, $v = 20\,\mathrm{m\,s^{-1}}$

So $35 \ln 20 = 0 + c$

So $c = 35 \ln 20$

$2x = 35 \ln 20 - 35 \ln v$

$x = \dfrac{35}{2} \ln \dfrac{20}{v}$

$\ln \dfrac{20}{v} = \dfrac{2}{35} x$

$\dfrac{20}{v} = e^{\frac{2}{35}x}$

$\dfrac{v}{20} = e^{-\frac{2}{35}x}$

$v = 20 e^{-\frac{2}{35}x}$

You want v in term of x, so rearrange the formula.

When $x = 10$:

$v = 20 e^{-\frac{2}{35} \times 10}$

$v = 11.3$

So after $10\,\mathrm{m}$, the ball will be travelling with a velocity of $11.3\,\mathrm{m\,s^{-1}}$.

Exercise 5.2A

1 A small sphere of mass $3\,\mathrm{kg}$ moves in a straight horizontal line. There is a resistance force of magnitude $21v\,\mathrm{N}$ acting on the sphere, where v is its speed. The sphere is released with a speed of $30\,\mathrm{m\,s^{-1}}$. Find the time taken for this speed to reduce by 50%.

2 A small ball of mass $1\,\mathrm{kg}$ is released from point A with a speed of $20\,\mathrm{m\,s^{-1}}$. There is a resistive force of $15v\,\mathrm{N}$ acting on the ball. Find the displacement from A when the speed becomes 25% of its original value.

3 A battery-powered toy car of mass $500\,\mathrm{g}$ has a driving force of $\dfrac{500}{v}\,\mathrm{N}$ and a resistance force of $4v$ acting on it. When $t = 0$, the car is moving at $10\,\mathrm{m\,s^{-1}}$. Find the speed of the car after 5 seconds.

4 A car of mass $800\,\mathrm{kg}$ leaves a village at a speed of $15\,\mathrm{m\,s^{-1}}$. The car has a driving force of $\dfrac{1600}{v}\,\mathrm{N}$ where v is the speed of the car in $\mathrm{m\,s^{-1}}$. The total resistance against the motion is $4v\,\mathrm{N}$.

   a Convert $70\,\mathrm{km\,h^{-1}}$ into $\mathrm{m\,s^{-1}}$.

   b Find the time taken for the car to increase its speed to $70\,\mathrm{km\,h^{-1}}$.

5 A car of mass $1000\,\mathrm{kg}$ is travelling along a horizontal road. The engine is giving the car a driving force of $\dfrac{4900}{v}\,\mathrm{N}$ where v is the velocity of the car and the magnitude of the resistance force acting on the car is $0.4v^2$. The speed of the car increases from $5\,\mathrm{m\,s^{-1}}$ to $12\,\mathrm{m\,s^{-1}}$. Find the distance travelled by the car during this increase in speed.

PS **6** A small object of mass 20 g is projected along a horizontal surface with a speed of $10\,\mathrm{m\,s^{-1}}$. The object is subject to a frictional force and a resistance force of size $5v^2$ where v is the speed of the object. The coefficient of friction between the object and the surface is 0.6. Find:

  a the magnitude of the frictional force acting on the object

  b the distance the object moves before its velocity reduces by 50%.

MM **7** In a competition, two men pull a trailer of mass 450 kg from rest to a speed of $4\,\mathrm{m\,s^{-1}}$ by applying a force of magnitude $\dfrac{45}{v+3}$ N. Air resistance is negligible. Find:

  a the distance moved while the speed increases from rest to $4\,\mathrm{m\,s^{-1}}$

  b the time taken for the speed to increase from rest to $4\,\mathrm{m\,s^{-1}}$.

Mathematics in life and work: Group discussion

You are working for a company that organises adventure holidays for groups of friends. The first activity involves racing quad bikes along an old airport runway. The manager of the company wants to ensure that if the brakes fail, there is enough runway left after the finish line for the quad bikes to come to a complete stop. He gives you the following information about the resistive force of three different models of quad bike.

Model	Total resistance force (N)
A	$400 + 4v^2$
B	$500 + 3v^2$
C	$900 + 70v$

The speed limit on the track is set at $30\,\mathrm{km\,h^{-1}}$ but the manager knows that some people will break this limit by up to $5\,\mathrm{km\,h^{-1}}$. You have been asked to design the track so that if the brakes do fail, there is sufficient distance for the quad bikes to stop.

1 The runway is 1.8 km in length. Where would you suggest the finish line is placed so that you meet the manager's safety requirement and give the maximum length of racetrack.

2 What other factors may affect the stopping distance of the quad bikes?

3 What would be the effect of increasing the speed limit on the maximum possible length of the track?

5.3 Vertical motion with air resistance

In Chapter 1, you studied objects being projected vertically upwards. To simplify the model, you used the assumption that air resistance is negligible. In this section, you will learn how to add resistance forces into the model to allow more accurate predictions to be made.

When you are modelling situations involving resistance, it is important to remember that the resistance force always opposes the direction of motion. This means that when an object is moving upwards, the resistance force is acting downwards and when the object is moving downwards, the resistance force is acting upwards. As a result, the upward and downward journeys need to be considered separately.

The techniques used in this section are similar to those used in previous sections. You need to find the resultant force acting on the object and use Newton's second law. If you need to find a time, use the relationship $a = \dfrac{dv}{dt}$. If you need to find a distance, use the relationship $a = v\dfrac{dv}{dx}$. This will give you a differential equation that you can solve.

Example 5

A ball of mass 0.75 kg is projected vertically upwards from the ground with a velocity of $15\,\text{m}\,\text{s}^{-1}$. The air resistance acting on the ball has a magnitude of $0.5v$ N where $v\,\text{m}\,\text{s}^{-1}$ is the speed of the object. Find:

a the greatest height of the ball

b the time taken for the ball to reach this maximum height.

Solution

a When the ball is moving upwards, there are two resistance forces on it that you need to consider: its weight and the air resistance. This gives you a total resistance of $(mg + 0.5v)$ N. This force is acting downwards so it will be negative.

Using Newton's second law:

$$ma = -mg - 0.5v$$

$$0.75a = -7.5 - 0.5v$$

$$3a = -30 - 2v$$

Divide both sides by 0.25.

$$3v\frac{dv}{dx} = -30 - 2v$$

$$\int \frac{3v}{2v + 30}\,dv = \int -1\,dx$$

The left-hand integral is an improper function so you need to divide $3v$ by $2v + 30$ before you integrate.

$$\int \frac{3}{2} - \frac{45}{2v + 30}\,dv = \int -1\,dx$$

$$\frac{3}{2}\int 1 - \frac{15}{v + 15}\,dv = \int -1\,dx$$

$$\frac{3}{2}(v - 15\ln(v + 15)) = -x + c$$

When $x = 0$, $v = 15$:

$$\frac{3}{2}(15 - 15\ln(15 + 15)) = -x + c$$

$$c = \frac{45}{2}(1 - \ln 30)$$

So $x = -\dfrac{3}{2}(v - 15\ln(v + 15)) - \dfrac{45}{2}(1 - \ln 30)$

You could substitute $v = 0$ into this equation without simplifying the right-hand side.

$$x = -\frac{3}{2}(v - 15\ln(v + 15) - 15 + 15\ln 30)$$

$$x = \frac{3}{2}\left(15\ln\left(\frac{v+15}{30}\right) - v + 15\right)$$

The maximum height of the ball is when $v = 0\,\mathrm{m\,s^{-1}}$.

Because the ball will instantaneously come to rest before changing direction to come back down.

$$x = \frac{3}{2}\left(15\ln\left(\frac{0+15}{30}\right) - 0 + 15\right)$$

$$= 6.90$$

The maximum height of the ball is 6.90 m.

b This part of the question is asking you to find a time, so when forming the differential equation you need to replace a by $\frac{\mathrm{d}v}{\mathrm{d}t}$.

From part **a**:

$$3a = -30 - 2v$$

$$3\frac{\mathrm{d}v}{\mathrm{d}t} = -30 - 2v$$

$$\int \frac{3}{2v+30}\,\mathrm{d}v = \int -1\,\mathrm{d}t$$

$$\frac{3}{2}\int \frac{1}{v+15}\,\mathrm{d}v = \int -1\,\mathrm{d}t$$

$$\frac{3}{2}\ln(v+15) = -t + c$$

When $t = 0$, $v = 15$

$$\frac{3}{2}\ln(15+15) = 0 + c$$

So $c = \frac{3}{2}\ln(30)$

This gives:

$$t = \frac{3}{2}\ln(30) - \frac{3}{2}\ln(v+15)$$

$$t = \frac{3}{2}\ln\left(\frac{30}{v+15}\right)$$

To find the time for the ball to reach its maximum height, you need to substitute $v = 0$ into the above formula.

$$t = \frac{3}{2}\ln\left(\frac{30}{0+15}\right) = 1.04$$

So the time taken for the ball to reach the maximum height is 1.04 seconds.

Stop and think Would the time taken to return to the ground be the same as the time taken to reach the maximum height?

When an object is dropped vertically, the weight of the object will be acting downwards and the air resistance will be acting upwards. Initially there will be a resultant force acting downwards. As the object gains speed, the air resistance acting on it will increase. Eventually, the magnitude of the air resistance and the weight will be the same, so there will be no resultant force acting on the object. This means that the object will continue to fall at a steady speed. This speed is called its **terminal velocity**.

In the next example, you will see two different ways of calculating the terminal velocity of an object falling vertically.

> **KEY INFORMATION**
>
> An object moving vertically will reach its terminal velocity when the resultant force acting on it is zero or its acceleration is zero.

Example 6

A skydiver of mass 80 kg falls out of an aircraft from rest. During her fall, she faces an air resistance of magnitude $20v$ N. Find her terminal velocity.

Solution

The resultant force acting on the skydiver will be $(mg - 20v)$ N = $(800 - 20v)$ N.

Using Newton's second law:

$ma = F$

$80a = 800 - 20v$

$4a = 40 - v$

You want to know the terminal velocity of the skydiver. To do this you can find an expression for the velocity in terms of time and explore the behaviour of the function as t tends to infinity. This means that you need to replace a by $\frac{dv}{dt}$.

$$4\frac{dv}{dt} = 40 - v$$

$$4\int \frac{1}{40 - v}\, dv = \int 1\, dt$$

$$-4\int \frac{-1}{40 - v}\, dv = \int 1\, dt$$

$$-4\ln(40 - v) = t + c$$

When $t = 0$, $v = 0$

$$-4\ln(40 - 0) = 0 + c$$

So $c = -4\ln 40$

So $t = 4\ln 40 - 4\ln(40 - v)$

$$4\ln\left(\frac{40}{40 - v}\right) = t$$

$$\ln\left(\frac{40}{40 - v}\right) = \frac{1}{4}t$$

By rearranging to get -1 as the numerator, you have a function in the form $\frac{f'(v)}{f(v)}$.

You need to make v the subject of the formula.

$$\frac{40}{40 - v} = e^{\frac{1}{4}t}$$

$$\frac{40 - v}{40} = e^{-\frac{1}{4}t}$$

$$40 - v = 40e^{-\frac{1}{4}t}$$

$$v = 40 - 40e^{-\frac{1}{4}t}$$

$$v = 40\left(1 - e^{-\frac{1}{4}t}\right)$$

As $t \to \infty$, $e^{-\frac{1}{4}t} \to 0$, $v \to 40$

So the terminal velocity of the skydiver is $40\,\mathrm{m\,s^{-1}}$.

An alternative and quicker method is to look at the differential equation $4\frac{dv}{dt} = 40 - v$ and substitute 0 for the acceleration,

that is, $\frac{dv}{dt} = 0$.

> When the resultant force is zero, the acceleration is $0\,\mathrm{m\,s^{-2}}$.

$$4\frac{dv}{dt} = 40 - v$$

$$0 = 40 - v$$

$$v = 40$$

So the terminal velocity of the skydiver is $40\,\mathrm{m\,s^{-1}}$.

Exercise 5.3A

1 A small ball of mass 300 g is dropped from rest into a liquid. The resistance force acting upon it has a magnitude of $2 + 0.5v^2$ N, where v is its velocity. Find the terminal velocity of the ball.

2 A small ball of mass 0.5 kg is projected vertically upwards with a velocity of $20\,\mathrm{m\,s^{-1}}$. The air resistance acting on the ball is $0.6v$ N, where v is the velocity of the ball. Find the maximum height the ball reaches.

3 A small object of mass 0.8 kg is projected vertically upwards with a speed of $15\,\mathrm{m\,s^{-1}}$. The air resistance acting on the object has magnitude $0.5v$. Find the time taken for the speed to reduce to $5\,\mathrm{m\,s^{-1}}$.

4 A particle of mass 0.3 kg is projected vertically upwards with a velocity of $10\,\mathrm{m\,s^{-1}}$. Find the maximum height reached by the particle if the air resistance acting on it has magnitude $0.5v^2$, where v is the speed of the particle.

5 A small package of mass 7 kg is dropped from an aeroplane at rest. While falling, the air resistance acting on it has magnitude $20v$ N. Find:

a its speed after falling for 3 seconds

b its terminal velocity.

6 A small object has a mass of 1 kg. It leaves the ground moving vertically upwards with a velocity of $25\,\mathrm{m\,s^{-1}}$. The resistance force acting on the object has magnitude $2v^2$. Find the time taken for the object to reach its maximum height.

7 A parachutist of mass 75 kg leaves an aircraft from rest. She freefalls for 4 seconds with an air resistance of magnitude of $25v$ N acting upon her. She then opens her parachute and has a resistance force of $20v + 40v^2$ acting on her. Find:

  a her speed after 4 seconds

  b her terminal velocity after her parachute has opened.

Mathematics in life and work: Group discussion

The package offered by the adventure holiday company includes skydiving. You have been asked to design a promotional leaflet for the company to explain the speeds reached during the dive. You find out that, during freefall, the air resistance acting on a person and their instructor, along with their kit is $30v$ N, and once the parachute is open, the resistance force acting on them is $(25v + 30v^2)$ N.

The skydiver will be attached to an instructor and released from a height of 3 km. They will then freefall for 45 seconds before the parachute is opened.

Assume that the average weight of a skydiver, the instructor and the kit is 200 kg.

1 What would the terminal velocity of the freefalling skydivers be if they fell for long enough?

2 Will the skydivers reach their terminal velocity before opening their parachute?

3 How far above the ground will they be when they open the parachute?

4 How long will it take them to reach the ground?

5 What other factors may affect the time taken for them to fall?

6 What would you suggest as the maximum time for which they should freefall before opening their parachute? Why?

SUMMARY OF KEY POINTS

› $a = \dfrac{dv}{dt} = \dfrac{d^2x}{dt^2}$

› $v = \dfrac{dx}{dt}$

› $v = \int a \, dt$

› $x = \int v \, dt$

› Using the chain rule, you can show that $a = v \dfrac{dv}{dx}$.

› A resistance force will act in the opposite direction to the direction of motion.

› If you need to write velocity in terms of time, replace a by $\dfrac{dv}{dt}$ to form a differential equation.

› If you need to write velocity in terms of displacement, replace a by $v\dfrac{dv}{dx}$ to form a differential equation.

› An object moving vertically will reach its terminal velocity when the resultant force acting on it is zero or its acceleration is zero.

EXAM-STYLE QUESTIONS

MM 1 A small sphere is moving along a straight horizontal line with an acceleration of magnitude $\dfrac{16}{(x+2)^2}$ m s^{-2} directed away from its starting point, where x is the distance away from the starting point. The sphere is moving with a speed of 0.5 m s^{-1} when $x = 2$ m.

    a Find v in terms of x.

    b Find the maximum value of v.

MM 2 A particle P of mass 0.4 kg is projected vertically upwards with a velocity of 25 m s^{-1}. The air resistance acting on P has magnitude $0.8v$ N, where v is the velocity of P at time t seconds.

    a Find the maximum height of P.

    b Find the value of t when P reaches its maximum height.

MM 3 A rock, P, of mass 2000 kg moves in a straight line on a horizontal surface. The velocity of P is v m s^{-1} when its displacement from O is x m. P has a driving force of $\dfrac{60\,000}{v}$ N and the magnitude of the resistance force acting on P is $5v^2$ N. The velocity of the rock at O is 4 m s^{-1}.

    a Find an expression for v in terms of x.

    b Find the distance travelled by P when its speed increases from 4 m s^{-1} to 15 m s^{-1}.

MM 4 A particle P moves in a straight line and passes through a fixed point O. The velocity of P at time t seconds is given by $v = e^{-5t} + 5t^2$. Initially, P is 3 m away from O.

    a Find the displacement of the particle from O when $t = 5$.

    b Find the magnitude of the acceleration of P when $t = 3$.

PS **5** A particle P moves along a straight horizontal line. Its acceleration in ms^{-2} at time t seconds is given by $a = \frac{1}{3}e^{-5t}$.

    a Given that P has an initial velocity of $5\,ms^{-1}$, find an expression for the velocity, $v\,ms^{-1}$, of P in terms of t.

    b Find the speed of P when $t = 3$.

    c Find the range of possible values of v.

MM **6** A small ball B of mass of $1\,kg$ is projected vertically upwards with a velocity of $25\,ms^{-1}$ from a height of $1.2\,m$ above the ground. The resistance force acting on B has magnitude $3v^2$.

    a Find the maximum height of B above the ground.

    b Find the time taken for B to reach its maximum height.

PS **7** A particle P is moving along a straight horizontal line from the point A to the point B. After t seconds, P has an acceleration of $\frac{1}{3}e^{-\frac{1}{2}x}\,ms^{-2}$ towards B and a velocity of $v\,ms^{-1}$ and a displacement of $x\,m$. P leaves A at a speed of $2\,ms^{-1}$ and $AB = 5\,m$.

    a Find an expression for v in terms of x.

    b Find the terminal velocity of P.

MM **8** A rock, P, of mass $1500\,kg$ starts from rest and moves along a horizontal straight line. P has a constant driving force of $9000\,N$ and encounters a resistance force of magnitude $(3000 + 350v)\,N$, where v is the velocity of P in ms^{-1} after t seconds.

    a Find an expression for v in terms of t.

    b Find the maximum velocity of P.

PS **9** A rock, P, of mass $50\,kg$ is released from rest at a height of $50\,m$ and falls vertically downwards. The air resistance acting on P is of magnitude $0.08v^2$ where v is the velocity of the rock in ms^{-1}. Calculate the speed of P when it reaches the ground.

MM **10** A particle P of mass $0.2\,kg$ is moving along a straight horizontal line from the point A. The displacement of P from A is $x\,m$ and its velocity is $v\,ms^{-1}$. The forces acting on P are $e^x\,N$ away from A and $(1 + 2x)\,N$ towards A. $AB = 10\,m$.

    a Show that $v\dfrac{dv}{dx} = 5e^x - 10x - 5$.

    b Given that P leaves A with a velocity of $4\,ms^{-1}$, find the speed of P when it arrives at B.

MM **11** A particle P of mass $m\,kg$ is initially at rest at a fixed point O before moving along a straight horizontal line. P accelerates away from O with a magnitude of $0.2e^{0.2x}\,ms^{-2}$ for 10 seconds, where x is the displacement from O. P then moves at a constant speed.

    a Find an expression for the velocity of P in terms of x.

    b Find the velocity of P when it is 10 meters from O.

    c Write down an expression for the acceleration of P after 10 seconds.

12 A particle P of mass 6 kg moves along a straight horizontal line. At time t seconds, the resultant force acting on P has magnitude $\frac{120}{kv^2}$ N in the direction of motion, where $k > 0$, v is the velocity of P in m s^{-1} and x is the displacement of P in m from a fixed point O. Initially, P is at O and is moving at 1 m s^{-1} and then, when $t = 1$ s, $v = 2$ m s^{-1}.

a Show that $k = \frac{60}{7}$.

b Show that $v^3 = 7t + 1$.

c Find an expression for v^4 in terms of x.

(PS) 13 A particle P of mass 60 kg moves along a straight horizontal line. At time t seconds, the resultant force acting on P has magnitude $\frac{10 \operatorname{cosec} v}{\cos v}$ N in the direction of motion, where v is the velocity of P in m s^{-1} and $0 < v < \frac{\pi}{2}$. The displacement of P from a fixed point O is x m. When $t = 0$, $v = \frac{\pi}{6}$.

a Show that $\cos 2v = \frac{3 - 4t}{6}$.

b Find the speed of P when $t = 1$.

(PS) 14 A particle P of mass 0.6 kg is moving along a straight horizontal line from the fixed point A. After t seconds, the displacement of P from A is x m and its velocity is v m s^{-1}. The forces acting on P are $(0.5v^2 + 1)$ N away from A and $(0.25v^2 + 0.5)$ N towards A. When P is at A, $v = \sqrt{2}$ m s^{-1}.

a Find an expression for v in terms of x.

b Find the speed of P when $x = 1.5$ m.

15 A particle P of mass 4 kg is moving along a straight horizontal line from a fixed point A. After t seconds, the displacement of P from A is x m and its velocity is v m s^{-1}. For $v > 0$ and $0 \leqslant x \leqslant 1$, the resultant force acting on P is $v^3 e^{5x-4}$ in the direction of motion. When P is 0.8 m away from A, $v = 2$ m s^{-1}.

a Show that $v = \frac{20}{11 - e^{5x-4}}$.

b Find the speed of P when $x = 0.5$ m.

(MM) 16 A parachutist of mass 80 kg leaves an aircraft from rest. He freefalls for 5 seconds with an air resistance of magnitude of $30v$ N acting upon him. He then opens his parachute and has a resistance force of $30v + 45v^2$ acting on him. By modelling the parachutist as a particle:

a find the speed of the parachutist at the instant the parachute opens

b find the terminal velocity of the parachutist after he has opened his parachute.

(PS) 17 A particle P of mass m kg is projected vertically upwards with a speed of 40λ m s^{-1} where λ is a positive constant. P encounters a resistance force of $\frac{mv^2}{4\lambda^2}$ N.

a Show that the maximum height reached by P is $2\lambda^2 \ln 41$ m.

b Find the time taken for P to reach its maximum height.

18 A particle P of mass m kg is projected vertically upwards with a speed of u m s^{-1} and is acted on by an air resistance of magnitude mkv^2, where v is its velocity in m s^{-1} at a given time, t seconds, and k is a positive constant.

   a Find the greatest height of P in terms of k, u and g.

   b Show that the time taken for P to reach this height is given by $t = \dfrac{1}{\sqrt{gk}}\tan^{-1}\left(u\sqrt{\dfrac{k}{g}}\right)$.

19 A particle P of mass 0.5 kg is moving along a straight horizontal line from a fixed point O.

   After t seconds, the displacement of P from A is x m and its velocity is v m s^{-1}. For $0 \leqslant x < \dfrac{\pi}{2}$,

   the resultant force acting on P is $\dfrac{x}{\sec(x^2)}$ N in the direction of motion. When P is at O, $v = 4$ m s^{-1}.

   a Show that $v^2 = \int (4x\cos(x^2))\,dx$.

   b Using the substitution $u = x^2$, or otherwise, find an expression for v in terms of x.

   c Find the range of possible values for v.

20 A particle P of mass 10 kg is moving along a straight horizontal line from a fixed point O. After t seconds, the displacement of P from A is x m and its velocity is v m s^{-1}. The resultant force acting on P is $x^2 e^{-v}$ N in the direction of motion. At O, the particle is at rest.

   a Show that $x^3 = 30\int (ve^v)\,dv$.

   b Using integration by parts, or otherwise, find an expression for x^3 in terms of v.

   c Find the displacement of P when $v = 2$ m s^{-1}.

Mathematics in life and work

A customer of an adventure company wants to know more about the skydiving activity. She wants to know what her maximum speed will be as she freefalls and at what speed she will land. The mass of the person, instructor and kit will be 180 kg. For the size of the kit you are using, you know that during freefall the total air resistance acting on a person, their instructor and their kit is $25v$ N and once the parachute is open the resistance force acting on them is $(20v + 28v^2)$ N.

1 The customer will freefall for 45 seconds before opening her parachute. What will her speed be when the instructor opens the parachute?

2 They leave the aeroplane at a height of 3200 m. How far above the ground will they be when the instructor opens the parachute?

3 What is the maximum speed at which they could land with the parachute open?

6 MOMENTUM

Mathematics in life and work

In this chapter, you will extend your knowledge about momentum from Mechanics 1 Chapter 3. You will look at Newton's experimental law, which allows you to predict what will happen when two objects collide. This technique is important in careers in the automotive industry and in sport – for example:

› If you were designing safety features of cars involving airbags, you would need to understand what happens after collisions at different angles to work out the best positions to place airbags.

› If you were a snooker player, you would need to understand the effect of Newton's experimental law when two balls collide or when one ball hits the cushion.

› If you were working with a theatre company that produces shows involving ball tricks, you would need to be able to calculate the different properties of the balls to predict what would happen.

LEARNING OBJECTIVES

You will learn how to:

› recall Newton's experimental law and definition of the coefficient of restitution, the property $0 \leq e \leq 1$, and the meaning of the terms 'perfectly elastic' and 'inelastic'

› use conservation of linear momentum and/or Newton's experimental law to solve problems that may be modelled as the direct or oblique impact of two smooth spheres or the direct or oblique impact of a smooth sphere with a fixed surface.

LANGUAGE OF MATHEMATICS

Key words and phrases you will meet in this chapter:

› coefficient of restitution, inelastic, Newton's experimental law, oblique impact, perfectly elastic

PREREQUISITE KNOWLEDGE

You should already know how to:

› use the definition of linear momentum and show understanding of its vector nature

› use conservation of linear momentum to solve problems that may be modelled as the direct impact of two bodies.

› solve a pair of simultaneous equations.

You should be able to complete the following questions correctly:

1 Find the momentum of an object of mass 2.6 kg moving at $3 \, \text{m s}^{-1}$ in a straight line.

2 Find the momentum of an object of mass 560 g moving with velocity $2 \, \text{m s}^{-1}$.

3 An object of mass 2 kg is moving in a straight line, with a speed of $5 \, \text{m s}^{-1}$. It collides with an object of mass 3 kg moving at a speed of $2 \, \text{m s}^{-1}$ in the same direction. Immediately after the collision the 3 kg object is moving at a speed of $4 \, \text{m s}^{-1}$ in the same direction. What is the speed and direction of the other object?

4 An object of mass 4 kg is moving in a straight line with a speed of $2 \, \text{m s}^{-1}$. It collides with an object of mass 4.5 kg moving at a speed of $3 \, \text{m s}^{-1}$ in the opposite direction. Immediately after the collision the 4 kg object is moving at a speed of $1 \, \text{m s}^{-1}$ in the opposite direction. What is the speed and direction of the other object?

5 Solve the following pairs of simultaneous equations.

  a $x + y = 5$
     $x - y = -1$

  b $2x + y = 1.3$
     $3x - y = 1.2$

6.1 Newton's experimental law and collisions with a stationary surface

In this section, you will extend your knowledge of linear momentum to work out the speed of two objects after a collision. Previously, in Mechanics 1, Chapter 3, you could work out the speed of one of the objects provided you knew the speed of the other.

When two objects collide, their speed after a collision depends on the material they are made from – some materials are more 'bouncy' than others.

Newton's experimental law relates the speed of two objects after a collision with the speeds before the collision and the properties of the objects. The **coefficient of restitution**, e, is defined as:

$$e = \frac{\text{speed of separation of particles}}{\text{speed of approach of particles}}$$

The coefficient of restitution can take values between 0 and 1 inclusive and does not have units.

If an object is very bouncy then e is very close to 1. If an object is not bouncy then e is close to 0. If $e = 1$, the collision is described as **perfectly elastic**. If $e = 0$, the collision is described as **inelastic**.

Kinetic energy is only conserved when a collision is perfectly elastic.

> **KEY INFORMATION**
>
> e = coefficient of restitution
>
> $0 \leqslant e \leqslant 1$
>
> If $e = 0$, the collision is inelastic.
>
> If $e = 1$, the collision is perfectly elastic and kinetic energy is conserved.

Collisions with a flat, smooth surface

You will now look at what happens when a moving object hits a surface at 90°. In order to keep the model simple, assume that the surface is smooth, flat and stationary.

Since the surface is stationary, the speed of approach is the same as the velocity of the object when it hits the surface and the speed of separation is the same as the velocity of the object when it leaves the surface.

Assume that the object hits the surface with velocity u m s^{-1} and leaves the surface with velocity v m s^{-1}.

Speed of approach = u and speed of separation = v

so $\dfrac{v}{u} = e$ or $v = eu$.

This relationship can be used alongside the equations of motion to answer questions about what happens after the impact.

Example 1

A small ball is dropped vertically and hits the ground with a speed of 4 m s^{-1}. The coefficient of restitution is 0.7. How high will the ball reach after impact?

Solution

First, you need to find the speed immediately after impact.

Since the ground is stationary:

$v = eu$

$\quad = 0.7 \times 4$

$\quad = 2.8$ m s^{-1}

Considering the upward motion:

$u = 2.8$ m s^{-1}

$v = 0$ m s^{-1} •⟶ This is the speed when it reaches its maximum height.

$a = -10$ m s^{-2}

$s = ?$

Using $v^2 = u^2 + 2as$:

$\qquad 0^2 = 2.8^2 + 2 \times -10 \times s$

$\qquad 20s = 7.84$

$\qquad\quad s = 0.392$

The ball will reach a maximum height of 39.2 cm after impact.

Mathematics in life and work: Group discussion

You are working as a special effects manager for a children's entertainment company. You are required to provide suitable props for the shows. As part of one of the shows, the main actors will do tricks with balls and they ask you to assist them in planning their act. They will use four different types of ball.

1 For one of their acts, the actors need to know how high each ball will bounce after they have dropped it. In order to do this, you need to calculate the coefficient of restitution between each ball and the stage. Design a simple experiment that will allow you to calculate the coefficient of restitution for each ball. Test this experiment on a selection of balls.

2 In one of the tricks, an actor on stilts will drop a ball vertically downwards from rest at a height of 2 m and wants it to bounce back up so that another actor can catch it with his hand, which is at a height of 1.2 m. What is the minimum coefficient of restitution between the ball and the surface for this to happen?

3 What other factors may affect the height the ball reaches?

Exercise 6.1A

(PS) 1 A ball is thrown horizontally at a vertical wall with a speed of $4 \, \text{m s}^{-1}$. After impact, it has a speed of $2.6 \, \text{m s}^{-1}$. What is the coefficient of restitution between the ball and the wall?

(PS) 2 A squash ball hits a wall with a speed of $12 \, \text{m s}^{-1}$ and rebounds with a speed of $10 \, \text{m s}^{-1}$. Find the coefficient of restitution.

(PS) 3 The coefficient of restitution between a ball and a wall is 0.55. The ball is thrown horizontally and rebounds off the wall with a speed of $3.2 \, \text{m s}^{-1}$. Find the speed at which the ball was thrown.

(PS) 4 A ball is dropped vertically from rest at a height of 1.5 m. The coefficient of restitution between the ball and the ground is 0.4. Find:

  a the speed of the ball when it hits the ground

  b the speed of the ball after it has hit the surface

  c the maximum height of the ball after impact.

(PS) 5 A ball is dropped vertically from a height of 1.42 m and reaches a maximum height of 0.75 m after impact. Find the coefficient of restitution between the ball and the surface.

(PS) 6 A ball is dropped from rest and bounces to a height of 2.1 m after impact. The coefficient of restitution is 0.76. Find the height from which the ball was released.

(MM) 7 A ball is dropped from a height of 3 m and is allowed to bounce until it comes to rest. The coefficient of restitution between the ball and the ground is 0.7. Find the maximum height reached after its third bounce.

c) 8 A ball is thrown horizontally at a wall with a speed of $15\,\mathrm{m\,s^{-1}}$ from a height of $1.2\,\mathrm{m}$. The coefficient of restitution between the ball and the wall is 0.6.

a Find the speed of the ball after impact.

b Calculate, showing all of your calculations, the time taken between the ball hitting the wall and hitting the ground for the first time.

c Find the distance between the wall and the point where the ball first hits the ground.

PS 9 A bouncy ball is thrown horizontally towards a wall at a height of $2\,\mathrm{m}$. After impact, the ball travels a distance of $5\,\mathrm{m}$ before it hits the ground. The coefficient of restitution between the ball and the wall is 0.8. Find the speed at which the ball was thrown.

6.2 Collisions between two spheres moving in a straight line

In the previous section, you looked at what happened when a moving object hit a stationary surface. In this section, you will consider what happens when a moving object hits another object moving in the same or the opposite direction.

Let u_1 and u_2 be the velocities of objects 1 and 2 before the collision and let v_1 and v_2 be their velocities after the collision. Then:

» the **speed of approach** is $u_1 - u_2$

» the **speed of separation** is $v_2 - v_1$.

This means that you can summarise Newton's experimental law as:

$\dfrac{v_2 - v_1}{u_1 - u_2} = e$

which can be written as $v_2 - v_1 = e(u_1 - u_2)$.

In Mechanics 1 Chapter 3, you learned that the principle of conservation of linear momentum states that the total momentum before a collision equals the total momentum after a collision.

Let m_1 and m_2 be the masses of objects 1 and 2 and let u_1 and u_2 be the velocities of objects 1 and 2 before the collision. Let v_1 and v_2 be the velocities after the collision.

The principle of conservation of linear momentum can be summarised as:

$m_1 u_1 + m_2 u_2 = m_1 v_1 + m_2 v_2$

Momentum is measured in $\mathrm{kg\,m\,s^{-1}}$.

> For a collision to occur, you need $u_1 > u_2$.

> You may find it easier to remember this by writing it as:
> $\dfrac{v_1 - v_2}{u_1 - u_2} = -e.$

Stop and think In real life, not all momentum will be conserved. What assumption is involved in the model?

If you combine Newton's experimental law with the principle of conservation of momentum, you can analyse the velocities of each object before and after the collision, as shown in the next two examples.

Example 2

Two toy balls, one blue and one red, are on a smooth track. Both balls have mass 2 kg. Alexander pushes the blue ball along the track so that it hits the stationary red ball with a speed of $3\,\text{m s}^{-1}$. The coefficient of restitution between the two balls is 0.5. Find the velocity of each ball after the collision.

Solution

It is always useful to represent the information in a diagram.

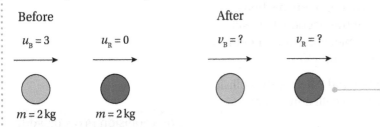

Before .... After

$u_B = 3$ ...... $u_R = 0$ ...... $v_B = ?$ ...... $v_R = ?$

$m = 2\,\text{kg}$ ...... $m = 2\,\text{kg}$

Speed of approach $= 3 - 0 = 3$

Speed of separation $= v_R - v_B$

By Newton's experimental law:

Speed of separation $= e \times$ speed of approach

$v_R - v_B = 0.5 \times 3$

$v_R - v_B = 1.5$ ①

By the conservation of linear momentum:

$m_B u_B + m_R u_R = m_B v_B + m_R v_R$

$2 \times 3 + 2 \times 0 = 2v_B + 2v_R$

$2v_B + 2v_R = 6$

$v_B + v_R = 3$ ②

You can now solve equations ① and ② simultaneously by adding them together.

$2v_R = 4.5$

$v_R = 2.25$

KEY INFORMATION

If objects of masses m_1 and m_2 and velocities u_1 and u_2 collide and have velocities of v_1 and v_2 after the collision, then:

› $v_2 - v_1 = e(u_1 - u_2)$

› $m_1 u_1 + m_2 u_2 = m_1 v_1 + m_2 v_2$

The model in this diagram assumes that both balls are moving towards the right after the collision. If this assumption is wrong, you will get a negative velocity, which will show that the ball is actually moving to the left.

Substitute this into equation ②:

$v_B + 2.25 = 3$

$v_B = 0.75$

So after the collision, the red ball has a velocity of $2.25\,\text{m s}^{-1}$ and the blue ball has a velocity of $0.75\,\text{m s}^{-1}$ in the same direction.

Example 3

In a children's ball game, a red ball is travelling with a velocity of $3\,\text{m s}^{-1}$ and collides with a black ball that is moving directly towards it with a speed of $0.5\,\text{m s}^{-1}$. If $e = 0.6$ and both balls have the same mass, describe what happens after the collision.

Solution

Before   After

    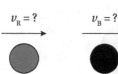

$u_R = 3$ $u_B = -0.5$ $v_R = ?$ $v_B = ?$

Speed of approach $= 3 - (-0.5) = 3.5$

Speed of separation $= v_B - v_R$

> v_B is negative because it is moving in the opposite direction to the red ball.

By Newton's experimental law:

Speed of separation $= e \times$ speed of approach

$v_B - v_R = 0.6 \times 3.5$

$v_R - v_B = 2.1$ ①

By the conservation of linear momentum:

$m_R u_R + m_B u_B = m_R v_R + m_B v_B$

$m \times 3 + m \times (-0.5) = mv_R + mv_B$

$mv_R + mv_B = 2.5m$

$v_B + v_R = 2.5$ ②

You can now solve equations ① and ② simultaneously by adding them together.

$2v_R = 4.6$

$v_R = 2.3$

Substitute this into equation ②:

$v_B + 2.3 = 2.5$

$v_B = 0.2$

So the black ball has a velocity of $2.5\,\text{m}\,\text{s}^{-1}$ and the red ball has a velocity of $0.2\,\text{m}\,\text{s}^{-1}$ in the same direction after the collision.

Although the examples have only looked at one collision taking place, the same rules can be used to look at multiple collisions. In cases where more than one collision takes place, you need to consider each collision separately.

Exercise 6.2A

(PS) 1 In each of the following cases, find the velocity of both balls after the collision.

a
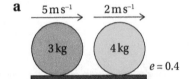

b
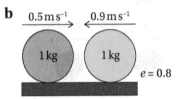

c
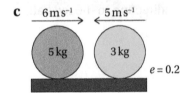

d
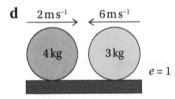

(PS) 2 A smooth sphere, A, of mass $2\,\text{kg}$ is moving along a straight line with a speed of $3\,\text{m}\,\text{s}^{-1}$ and collides with a second smooth sphere, B, of mass $3\,\text{kg}$ which is stationary. The coefficient of restitution between the two spheres is 0.7. Find the speed and direction of sphere B after the collision.

(C) 3 Two balls A and B have mass $4m$ and $5m$, respectively. The balls are rolled towards each other on a smooth track, each with a speed of $5\,\text{m}\,\text{s}^{-1}$. The coefficient of restitution between the two balls is 0.55. Describe what happens to each ball after the collision.

(MM) 4 A bowling ball of mass $4\,\text{kg}$ is rolled at a speed of $5\,\text{m}\,\text{s}^{-1}$ along a straight line towards another ball of mass $3\,\text{kg}$, which is stationary. The coefficient of restitution between the two balls is 0.6. Find the velocities of both balls after the collision.

5 Two small spheres, A and B, of mass $2m$ and $3m$, respectively, are moving towards each other in a straight line. The speed of A is u ms^{-1}, the speed of B is $3u$ ms^{-1} and the coefficient of restitution between A and B is e.

a Show that $v_B = \frac{u}{5}(8e - 7)$.

b Find the velocity of sphere A.

 6 Three balls, A, B and C, all of mass m, are stationary on a smooth straight track. The coefficient of friction between any two balls is 0.6. Ball A is pushed towards ball B with a speed of 4 ms^{-1}. After impact, ball B moves towards ball C and collides with it. Find the speed of ball C after this collision.

6.3 Oblique impact of a smooth sphere with a fixed surface

In Section 6.1, you learned how to solve problems involving smooth spheres or spheres colliding with a perpendicular, flat, smooth surface. In this section, you will solve problems involving collisions at different angles. This is called **oblique impact** or oblique collision.

The diagram below shows a sphere with speed U colliding with a fixed surface at an angle α above the surface. It then leaves the surface after impact with speed V at an angle β above the surface.

Before impact   After impact

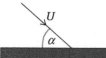   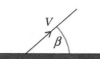

The speed can be resolved parallel and perpendicular to the surface. The component parallel to the surface will not be changed, but the component perpendicular to the surface will be affected by the coefficient of restitution between the sphere and the surface. If the speed perpendicular to the surface is $U \sin\alpha$ before impact, then the speed perpendicular to the surface after impact will be $eU \sin\alpha$, as shown in the diagram below.

Before impact   After impact
$U \sin\alpha$ $eU \sin\alpha$

If the sphere leaves the surface at an angle of β then:

$$\tan\beta = \frac{eU \sin\alpha}{U \cos\alpha}$$

$$\tan\beta = e \tan\alpha$$

> **KEY INFORMATION**
>
> If a sphere hits a surface at an angle α with a speed of u ms^{-1}, then after the collision:
>
> › the speed parallel to the surface $= u \cos\alpha$
>
> › the speed perpendicular to the surface $= eu \sin\alpha$
>
> where e is the coefficient of restitution between the sphere and the surface.

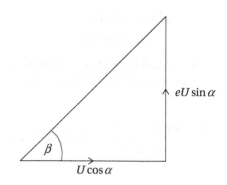

Example 4

A ball is thrown at the ground. It hits the ground with a velocity of $12\,\mathrm{m\,s^{-1}}$ at an angle of 40° above the horizontal. At what velocity does it leave the ground if $e = 0.5$?

Solution

Represent the information in the question as a diagram. Resolve the velocity into components parallel and perpendicular to the surface.

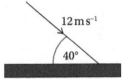

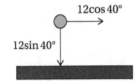

After the collision:

Parallel to the surface, the velocity will remain the same at $12\cos 40$.

Perpendicular to the surface, the velocity will be $0.5 \times 12\sin 40° = 6\sin 40°$.

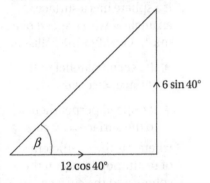

$$v = \sqrt{(12\cos 40°)^2 + (6\sin 40°)^2}$$

$$= 9.97\,\mathrm{m\,s^{-1}} \text{ to 3 s.f.}$$

$$\tan\beta = \frac{6\sin 40°}{12\cos 40°}$$

$$\beta = \tan^{-1} 0.420$$

$$= 22.8° \text{ to 1 d.p.}$$

So the ball leaves the ground with a velocity of $9.97\,\mathrm{m\,s^{-1}}$ at an angle of 22.8° above the surface.

Stop and think   What effect does the value of e have on the principle of the conservation of energy?

Example 5

Sally is playing snooker with her friend. The ball hits the cushion at a speed of $3\,\text{m s}^{-1}$ with an angle of $45°$ between the cushion and the direction of movement. The coefficient of restitution between the snooker ball and the cushion is 0.8. Find the velocity and direction of the ball after the impact.

Solution

Before the collision:

Parallel to the cushion, the velocity is $3\cos 45°\,\text{m s}^{-1}$.

Perpendicular to the cushion, the velocity is $3\sin 45°\,\text{m s}^{-1}$.

After the collision:

Parallel to the cushion, the velocity is $3\cos 45°\,\text{m s}^{-1}$.

Perpendicular to the cushion, the velocity is $0.8 \times 3\sin 45°\,\text{m s}^{-1}$.

So the velocity after impact is:

$$v = \sqrt{(3\cos 45°)^2 + (0.8 \times 3\sin 45°)^2}$$

$$= 2.72\,\text{m s}^{-1} \text{ to 3 s.f.}$$

Let β be the angle between the cushion and ball after impact.

Then $\tan \beta = \dfrac{0.8 \times 3\sin 45°}{3\cos 45°}$

$$\beta = \tan^{-1} 0.8$$

$$= 38.7° \text{ to 1 d.p.}$$

So the snooker ball will leave the cushion with an angle of $38.7°$ between the cushion and the direction of movement, at a speed of $2.72\,\text{m s}^{-1}$.

Exercise 6.3A

(PS) **1** A particle hits a smooth horizontal surface at an angle of $30°$ above the horizontal at a speed of $4.5\,\text{m s}^{-1}$. The coefficient of restitution between the particle and surface is 0.45. Find the velocity of the particle after impact.

(PS) **2** A football is kicked and hits the ground at an angle of $50°$ above the horizontal with a velocity of $12\,\text{m s}^{-1}$. The coefficient of restitution between the ball and the ground is 0.4. Assuming that the ground and ball are smooth, find the speed and direction of the ball immediately after impact.

(MM) **3** A small ball is thrown against a wall so it hits the wall with a speed of $2.5\,\text{m s}^{-1}$ at an angle of $80°$ between the direction of the ball and the wall. The coefficient of restitution is 0.7. Find the speed and direction of the ball immediately after the impact.

(MM) **4** During a game of snooker a ball travelling at $4\,\text{m s}^{-1}$ hits the cushion with an angle of $72°$ between its direction of motion and the cushion. The coefficient of restitution between the ball and the cushion is 0.78. Find the speed and direction of the ball immediately after the collision.

(PS) **5** After impact, a particle leaves the surface at an angle of 30° with a velocity of 3.2 m s⁻¹. The coefficient of restitution is 0.3. Find:

    a the angle at which the particle hit the surface

    b the speed of the particle immediately before impact.

(PS) **6** A small ball hits the ground at an angle of 36° to the horizontal with a speed of 3.2 m s⁻¹. It leaves the ground with a speed of 2.8 m s⁻¹ at an angle α to the horizontal.

    a Find angle α.

    b Find the coefficient of restitution between the ball and the ground.

(MM) **7** Shu is playing a game of crazy golf. To avoid an obstacle and get the ball in the hole, she calculates that she needs to bounce the ball off a point on a wall so that it leaves the wall at an angle of 50°. She has calculated that the coefficient of restitution between the ball and the wall is 0.88 and that the ball will hit the wall with a speed of 6 m s⁻¹. What must the angle be between the ball and the wall at the point of impact for her shot to be successful?

(MM) **8** Amrit and Kadija are playing a game that involves bouncing a ball to each other. Amrit throws the ball from a height of 1 m and it hits the ground at an angle of 40° with a speed of 3.2 m s⁻¹.
(C) The coefficient of restitution between the ball and the ground is 0.6. Assuming that the ball and the ground are smooth, calculate:

    a the speed and direction of the ball immediately after impact

    b the maximum height of the ball after impact

    c the distance between the first bounce and the second bounce, assuming the ball is not caught.

(MM) **9** A smooth ball is released from a height of 2 m at a speed of 5.1 m s⁻¹ and an angle of 30° above the horizontal. The ball bounces once and then is caught when it reaches its maximum
(C) height. The coefficient of restitution between the ball and the ground is 0.7.

    a Find the horizontal distance from the point of release to the point where it is caught.

    b Find the time taken from release to being caught.

    c State any assumptions you have made in your model.

(MM) **10** A ball is thrown horizontally at a speed of 10 m s⁻¹ from a height of 5 m. The coefficient of restitution between the ball and the ground is 0.65. After how many bounces will the maximum height of the ball be less than 50 cm?

Mathematics in life and work: Group discussion

For the second act of the children's show, an actor is planning to impress the audience with his basketball skills. His plan is to score goals by bouncing a ball off another object before it hits a target board. You have been asked to provide him with some calculations to help him to perfect his tricks. For this trick, a special ball has been designed so that the coefficient of restitution between the ball and the surfaces is 0.65. The target board is 2 m above the ground.

1   For the first trick, the actor will stand at the top of a ladder that is 5 m tall. He will release the ball at an angle of 45° above the horizontal with a speed of $10\,\mathrm{m\,s^{-1}}$. He wants the ball to bounce on the ground and then hit the target board.

   a At what speed and angle will the ball first hit the ground?

   b What will be the speed and angle of the ball immediately after impact?

   c How long after the first bounce will the ball first reach 2 m?

   d What is the required horizontal distance between the ladder and the target board so that the ball hits the target board at 2 m?

   e What assumptions have you made?

2   For his second trick, he would like to hit the target while facing away from it. His aim is to bounce the ball off a wall in front of him so that it will then go directly from the wall to the target. Plan a set-up for the stage so that he knows how far away he needs to be away from the wall, how far the target needs to be behind him, and also the speed and angle at which he needs to release the ball. State any assumptions you make.

3   The actor has asked you to design his third trick. You need to provide him with a detailed plan explaining how to set up and carry out this trick.

6.4 Oblique impact of two smooth spheres

In this section, you will learn how to find the velocities of two smooth spheres when they collide together at different angles.

At the point when two smooth spheres collide, there is a common tangent that is perpendicular to the line joining the centres of the two spheres, as shown in the diagram below.

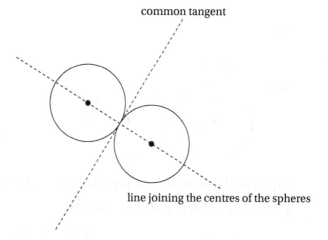

common tangent

line joining the centres of the spheres

When you solved problems involving oblique impacts with a smooth surface, you resolved the velocity parallel and perpendicular to the surface. In the case of two smooth spheres colliding, you need to resolve the velocity so that it is parallel and perpendicular to the line joining the centres.

As in the case of oblique impacts with a smooth surface, the component of the velocity perpendicular to the line joining the centres of the spheres will be unaffected by the collision. This means that, for both spheres, the components of the velocity perpendicular to this line will be the same before and after the collision. The components of the velocities parallel to this line may change after impact and can be calculated using the conservation of linear momentum and Newton's experimental law. This can be seen in the following examples.

Example 6

A smooth sphere, A, of mass 2 kg is moving towards another smooth sphere, B, of mass 4 kg, which is at rest. Just before impact, sphere A is travelling with a speed of 5 m s⁻¹ at an angle of 45° with the line joining the centres of the two spheres. The coefficient of restitution between the two spheres is 0.45. Find the velocities of both spheres directly after impact.

Solution

Present the information in a diagram.

Before

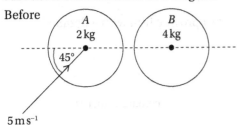

5 m s⁻¹

After

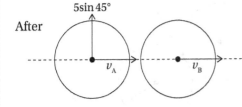

Perpendicular to the line joining the centres of the spheres:

Sphere A: speed after collision = $5\sin 45°$

$$= 3.54 \text{ m s}^{-1} \text{ to 3 s.f.}$$

Sphere B: speed after collision = 0 m s^{-1}

Parallel to the line joining the centres of the spheres:

By the conservation of linear momentum:

$$2 \times 5 \cos 45° + 4 \times 0 = 2 \times v_A + 4 \times v_B$$

$$2v_A + 4v_B = 10\cos 45°$$

$$v_A + 2v_B = 5\cos 45° \qquad ①$$

> **KEY INFORMATION**
>
> If two smooth spheres collide:
>
> › the components of the velocity of each sphere perpendicular to the line joining the centres of the spheres are unchanged
>
> › the components of the velocity of each sphere parallel to the line joining the centres of the spheres will change.

> Since sphere B is at rest and hence has no component of velocity perpendicular to the line joining the centres of the spheres, it will have no component of velocity in this direction after the impact.

By Newton's experimental law:

$$v_B - v_A = 0.45 \times 5\cos 45°$$

$$v_B - v_A = 2.25\cos 45° \qquad ②$$

You can now solve equations ① and ② simultaneously by adding them together.

$$3v_B = 7.25\cos 45°$$

$$v_B = 1.71\,\mathrm{m\,s^{-1}} \text{ to 3 s.f.}$$

Substitute this into equation ①:

$$v_A + 3.42 = 5\cos 45°$$

$$v_A = 0.116\,\mathrm{m\,s^{-1}} \text{ to 3 s.f.}$$

Sphere A:

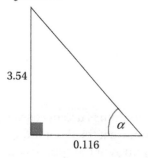

3.54

0.116

α

$$\text{Velocity after impact} = \sqrt{3.54^2 + 0.116^2}$$

$$= 3.54\,\mathrm{m\,s^{-1}}$$

$$\tan \alpha = \frac{3.54}{0.116}$$

$$\alpha = \tan^{-1}\left(\frac{3.54}{0.116}\right)$$

$$= 88.1°$$

So the final velocity of A is $3.54\,\mathrm{m\,s^{-1}}$ at an angle of $88.1°$ above the line joining the centres of the two spheres moving toward the top right of the diagram.

Sphere B is travelling at a velocity of $1.71\,\mathrm{m\,s^{-1}}$ parallel to the line joining the centres of the two spheres.

Example 7

Two snooker balls of the same mass are rolling towards each other. Ball A is travelling at $4\,\text{m s}^{-1}$ in a direction at 65° to the line joining the centres of the balls and ball B is travelling at $5\,\text{m s}^{-1}$ in a direction at 60° to the line joining the centres of the balls. The coefficient of restitution between the balls is 0.7. Describe what happens to the snooker balls after the collision.

Solution

The above scenario can be represented by the diagram below.

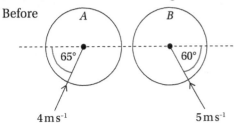

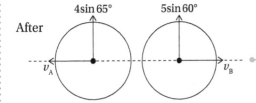

The problem is usually easier to visualise if the line joining the centres of the spheres is horizontal.

Perpendicular to the line joining the centres of the spheres:

Sphere A: speed after collision = $4\sin 65°$

$$= 3.63\,\text{m s}^{-1} \text{ to 3 s.f.}$$

Sphere B: speed after collision = $5\sin 60°$

$$= 4.33\,\text{m s}^{-1} \text{ to 3 s.f.}$$

Parallel to the line joining the centres of the spheres:

By the conservation of linear momentum:

$$m \times 4\cos 65 + m \times -5\cos 60 = m \times v_A + m \times v_B$$

$$v_A + v_B = 4\cos 65° - 5\cos 60°$$

$$v_A + v_B = -0.810 \qquad ①$$

v_B is negative because it is moving in the opposite direction to A.

By Newton's experimental law:

$$v_B - v_A = 0.7 \times (4\cos 65° - (-5\cos 60°))$$

$$v_B - v_A = 2.93 \qquad ②$$

You can now solve equations ① and ② simultaneously by adding them together.

$2v_B = 2.12$

$v_B = 1.06\,\text{m s}^{-1}$ to 3 s.f.

Substitute this into equation ①:

$v_A + 1.06 = -0.810$

$v_A = -1.87\,\text{m s}^{-1}$ to 3 s.f.

Sphere A:

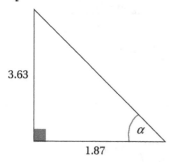

3.63

1.87

Velocity after impact $= \sqrt{3.63^2 + 1.87^2}$

$$= 4.08\,\text{m s}^{-1}$$

$$\tan\alpha = \frac{3.63}{1.87}$$

$$\alpha = \tan^{-1}\left(\frac{3.63}{1.87}\right)$$

$$= 62.7°$$

So the final velocity of A is $4.08\,\text{m s}^{-1}$ at an angle of 62.7° above the line joining the centres of the two spheres, away from B.

Sphere B:

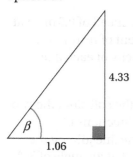

4.33

1.06

Velocity after impact $= \sqrt{4.33^2 + 1.06^2}$

$$= 4.46\,\text{m s}^{-1}$$

$$\tan\beta = \frac{4.33}{1.06}$$

$$\beta = \tan^{-1}\left(\frac{4.33}{1.06}\right)$$

$$= 76.2°$$

So the final velocity of B is $4.46\,\text{m s}^{-1}$ at an angle of 76.2° above the line joining the centres of the two spheres, away from A.

Exercise 6.4A

1 Find the magnitude and direction of the velocities of sphere A and sphere B, which have equal mass, immediately after each of these collisions.

a

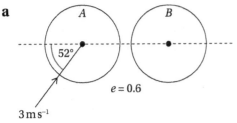

$e = 0.6$

$3\,\mathrm{m\,s^{-1}}$

b

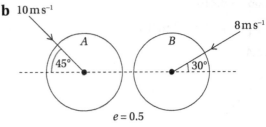

$e = 0.5$

c

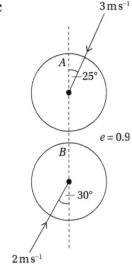

$e = 0.9$

$2\,\mathrm{m\,s^{-1}}$

d

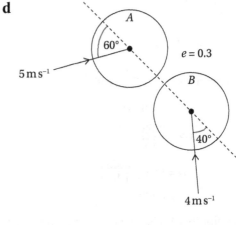

$e = 0.3$

$4\,\mathrm{m\,s^{-1}}$

2 A red pool ball collides with a stationary blue ball of the same mass with a speed of $0.5\,\mathrm{m\,s^{-1}}$ at an angle of $15°$ to the line joining the centres of the two balls. The coefficient of restitution between the two balls is 0.3. Find the magnitude and direction of the velocity of each ball directly after impact.

3 Two small marbles of the same mass collide obliquely. Immediately after the collision, the two marbles move away from each other with an angle of $125°$ between their directions of movement. The first marble has a velocity of $2\,\mathrm{m\,s^{-1}}$ at an angle of $25°$ to the line joining the centres of the two marbles and the second marble has a velocity of $1.5\,\mathrm{m\,s^{-1}}$ at an angle of $30°$ to the line joining the centres of the two marbles. The coefficient of restitution between the two marbles is 0.4. Find the magnitude and direction of the velocity of each marble before the collision.

4 Two small smooth spheres, A and B, of equal mass are moving with equal initial speeds, u, but in different directions. At the point of collision, A is moving parallel to the line joining the centres of the spheres and B is moving perpendicular to the line joining the centres of the two spheres. The coefficient of restitution between the two spheres is 0.5. Find the magnitude and direction of the velocity of each sphere after the collision.

5 Two uniform smooth spheres, *A* and *B*, of equal radius are moving in opposite directions on a smooth horizontal surface when they collide obliquely. The masses of *A* and *B* are 4 kg and 1 kg, respectively. Immediately before the collision the speed of *A* is 16 m s⁻¹ and the speed of *B* is 10 m s⁻¹. At the point of collision, the angle between the direction of the spheres and the line joining the centres of the spheres is θ where $\tan\theta = \frac{5}{4}$. Given that the coefficient of restitution between the two spheres is 0.22, determine the magnitude and direction of the velocities of *A* and *B* directly after the impact.

SUMMARY OF KEY POINTS

> The coefficient of restitution, $e = \dfrac{\text{speed of separation}}{\text{speed of approach}}$

> The coefficient of restitution can take values between 0 and 1 inclusive and does not have units.

> If $e = 1$ the collision is described as perfectly elastic. If $e = 0$ the collision is inelastic.

> Let u_1 and u_2 be the velocities of objects 1 and 2 before the collision and v_1 and v_2 be the velocities after the collision:

> > the speed of approach is $u_1 - u_2$

> > the speed of separation is $v_2 - v_1$

> > $\dfrac{v_2 - v_1}{u_1 - u_2} = e$ or $\dfrac{v_1 - v_2}{u_1 - u_2} = -e$.

> The principle of conservation of linear momentum states $m_1 u_1 + m_2 u_2 = m_1 v_1 + m_2 v_2$.

> If a sphere hits a surface at an angle α with a speed of u m s^{-1}, then, after the collision:

> > the speed parallel to the surface $= u \cos \alpha$

> > the speed perpendicular to the surface $= eu \sin \alpha$.

> If two smooth spheres collide:

> > the components of the velocity of each sphere perpendicular to the line joining the centres of the spheres are unchanged

> > the components of the velocity of each sphere parallel to the line joining the centres of the spheres will change.

EXAM-STYLE QUESTIONS

1 Two smooth spheres A and B, of equal radii and masses $5m$ and $3m$, respectively, lie at rest on a smooth horizontal table. The spheres A and B are projected directly towards each other with speeds $2u$ and $4u$, respectively. The coefficient of restitution between the spheres is e. Find the set of values of e for which the direction of motion of B is reversed in the collision.

2 A smooth sphere is moving along a horizontal surface at a constant speed of 5 m s^{-1} when it collides with a fixed vertical wall at a horizontal angle of $45°$ to the wall. Immediately after the impact the speed of the sphere is 4 m s^{-1}. Find the coefficient of restitution between the sphere and the wall.

3 A small smooth sphere S is moving on a smooth horizontal surface with speed 16 m s^{-1}. S collides with a smooth vertical barrier at a horizontal angle θ to the barrier. The coefficient of restitution between S and the barrier is 0.6. Given that the speed of S is $\frac{3}{4}$ of the original speed as a result of the collision, find the value of θ.

PS 4 A smooth particle of mass $3m$ is moving with speed u on a flat surface and collides directly with another particle of mass $5m$ that is stationary. The coefficient of restitution between the two particles is e. Find the speeds of A and B directly after the collision in terms of e and u.

PS **5** A small smooth ball hits the ground at an angle of 42° to the ground with a speed of $5.2\,\text{m}\,\text{s}^{-1}$. It leaves the ground with a speed of $3.9\,\text{m}\,\text{s}^{-1}$ at an angle α to the horizantal.

   a Find the coefficient of restitution between the ball and the ground.

   b Find the value of α.

PS **6** A smooth ball is dropped from a height of 2 m and bounces to a height of 1.2 m after it collides with a smooth horizontal surface.

   a Find the speed of the ball immediately before the collision.

   b Find the speed of the ball immediately after the collision.

   c Find the coefficient of restitution between the ball and the surface.

MM **7** A small smooth bouncy ball is dropped from rest at a height of 6 m on to a fixed flat horizontal surface. The coefficient of restitution between the ball and the surface is 0.8.

   a Find the maximum height of the ball after its first bounce.

   b Find the time that elapses between the first and second bounce.

   c Find the velocity of the ball immediately before its second impact.

MM **8** A smooth sphere is thrown at a solid vertical wall. It hits the wall horizontally with a speed of $12\,\text{m}\,\text{s}^{-1}$ at a height of 2 m. The coefficient of restitution between the wall and the ball is 0.7.

   a Find the speed of the ball immediately after the collision.

   b Find the time taken for the ball to reach the ground for the first time.

   c Find the time that elapses between the first and second bounce if the coefficient of restitution between the ball and the ground is 0.5.

C **9** Two smooth balls A and B, of equal radii and equal masses are on a smooth track, perpendicular to the line of a vertical wall. Ball A moves along the track at $4\,\text{m}\,\text{s}^{-1}$ and collides with ball B, which is stationary. In the subsequent motion, ball B bounces off the wall and collides once again with ball A. The coefficient of restitution between the two balls is 0.7 and between the balls and the wall is 0.5.

   a Find the speeds of ball A and ball B after the first collision.

   b Find the speed of B immediately before its second collision with A.

PS **10** A smooth sphere A of mass 50 g is travelling with a speed of $4\,\text{m}\,\text{s}^{-1}$ towards a stationary sphere B of mass 40 g. The spheres have equal radii and they collide in the direction of the line joining the centres of the two spheres. After the collision, sphere A is moving in the same direction with a speed of $1\,\text{m}\,\text{s}^{-1}$.

   a Find the velocity of sphere B after the collision.

   b Find the coefficient of restitution between the two spheres.

11 A particle P with mass m collides with a smooth vertical wall at $5u$ m s^{-1} at an angle of 30° to the wall. Immediately after the collision, P is travelling with speed λu m s^{-1}, where λ is a positive constant, at 90° to the original direction and in the same horizontal plane as the impact trajectory.

    a Find the coefficient of restitution between the wall and the particle.

    b Find the value of λ.

12 A small ball is thrown from a height of 3 m at an angle 40° above the horizontal with a speed of 4 m s^{-1}. It bounces on a smooth horizontal surface. The coefficient of restitution between the ball and surface is 0.6.

    a Find the speed and direction of the ball immediately after the first bounce.

    b Find the horizontal distance travelled between the ball being thrown and the second bounce.

13 A small smooth ball is dropped from rest from a height of 10 m on to a fixed horizontal surface. The coefficient of restitution between the ball and the surface is e. Show that the time taken for the ball to stop bouncing is $\sqrt{2} \times \dfrac{1+e}{1-e}$.

14 Two smooth spheres A and B, of equal radii and masses $2m$ and $3m$, respectively, collide obliquely. A is travelling at a speed of 3.5 m s^{-1} toward B at an angle of 35° to the line joining the centres of the spheres and B is travelling at a speed of 2.9 m s^{-1} at an angle of 40° to the line joining the centres of the spheres.

    a Find the velocities of A and B immediately after the impact.

After the first collision, sphere B then collides with a wall that is parallel to the line joining the centres of the spheres. The coefficient of restitution between the ball and the wall is 0.6.

    b Find the velocity of B after the collision with the wall.

15 Two smooth balls A and B, of equal radii and masses 3 kg and 4 kg, respectively, lie at rest on a smooth horizontal table. The coefficient of restitution between the two balls is 0.2. Ball A is travelling at 4 m s^{-1} and it collides directly with ball B, which is travelling at 2 m s^{-1} in the opposite direction. After the collision ball B travels off the edge of the table and bounces on the horizontal ground that is 0.8 m below.

    a Find the velocity of each ball immediately after the collision.

    b Find the distance away from the base of the table that the ball B lands on its first bounce.

16 A smooth sphere, A, is moving along a smooth horizontal plane when it collides obliquely with stationary sphere, B, with the same radius and mass. Immediately before the collision, sphere A is moving with speed u in a direction that makes an angle of 50° with the line joining the centres of the spheres. The coefficient of restitution between the two spheres is e.

    a Find, in terms of e and u, the velocity of A immediately after the collision.

    b Find, in terms of e and u, the velocity of B immediately after the collision.

17 Three uniform small smooth spheres A, B and C have equal radii and masses 5 kg, 3 kg and 1 kg, respectively. The spheres are at rest in a straight line on a smooth horizontal surface, with B between A and C. The coefficient of restitution between A and B is e and the coefficient of restitution between B and C is $2e$. Sphere A is projected directly towards B with speed $0.5\,\mathrm{m\,s^{-1}}$. In the case where $e = 0.4$, find the speed of C after B has collided with it.

18

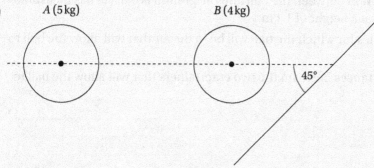

A (5 kg)   B (4 kg)   45°

A smooth sphere A of mass 5 kg is travelling with a speed of $4\,\mathrm{m\,s^{-1}}$ towards a stationary smooth sphere B of mass 4 kg. The spheres have equal radii and they collide in the direction of the line joining the centres of the two spheres. The coefficient of restitution between A and B is 0.8.

a Find the speeds of A and B after the collision.

After the initial collision, sphere B collides with a vertical wall at a horizontal angle of 45° to the wall. After sphere B collides with the wall, it travels with speed $2.9\,\mathrm{m\,s^{-1}}$ at an angle θ with the wall.

b Find the coefficient of restitution between sphere B and the wall.

c Find the value of θ.

19 Three uniform small smooth spheres, A, B and C, have equal radii and equal masses. They lie in a straight line on a smooth horizontal surface with B between A and C. Initially, A is moving towards B with speed $2\,\mathrm{m\,s^{-1}}$, and B and C are at rest. Sphere A collides and coalesces with sphere B to form the object AB. The object AB collides with sphere C and during the collision AB loses $\frac{2}{3}$ of its kinetic energy. The coefficient of restitution between the object AB and the sphere C is e. Find the value of e.

Mathematics in life and work

A children's entertainer is planning to carry out a trick involving a ball. He plans to throw a small ball so that it bounces on the ground and then lands in the mouth of his assistant who will be sitting in a chair. He will throw the ball horizontally from a height of 2 m with a speed of $10\,\text{m}\,\text{s}^{-1}$. The coefficient of restitution between the ball and the ground is 0.8. He has measured that his assistant's mouth will be at a height of 1.1 m.

1 Calculate the two possible times for which the ball will be in the air that will allow the ball to land in the assistant's mouth.

2 Calculate the two possible distances between the two entertainers that will allow the ball to land in the assistant's mouth.

SUMMARY REVIEW

Practise the key concepts and apply the skills and knowledge that you have learned in the book with these carefully selected past paper questions, supplemented with exam-style questions and extension questions written by the authors.

> Warm-up Questions

> A Level Questions

> Extension Questions

Three Cambridge International A Level Mathematics past paper questions based on prerequisite skills and concepts that are relevant to the main content of this book.

Selected past paper exam questions and exam-style questions on the topics covered in this syllabus component.

Extension questions that give you the opportunity to challenge yourself and prepare you for more advanced study.

Warm-up questions

Reproduced by permission of Cambridge Assessment International Education

1

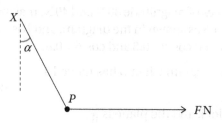

A particle P of mass $0.3\,\text{kg}$ is attached to one end of a light inextensible string. The other end of the string is attached to a fixed point X. A horizontal force of magnitude $F\,\text{N}$ is applied to the particle, which is in equilibrium when the string is at an angle α to the vertical, where $\tan\alpha = \dfrac{8}{15}$ (see diagram). Find the tension in the string and the value of F. **[4]**

Cambridge International AS & A Level Mathematics 9709 Paper 41 Q1 Nov 2013

2 A block is at rest on a rough horizontal plane. The coefficient of friction between the block and the plane is 1.25.

   i State, giving a reason for your answer, whether the minimum vertical force required to move the block is greater than or less than the minimum horizontal force to move the block. **[2]**

A horizontal force of continuously increasing magnitude P N and fixed direction is applied to the block.

   ii Given that the weight of the block is 60 N, find the value of P when the acceleration of the block is $4\,\text{m s}^{-2}$. **[2]**

Cambridge International AS & A Level Mathematics 9709 Paper 41 Q1 June 2013

3

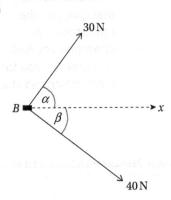

A block B lies on a rough horizontal plane. Horizontal forces of magnitude 30 N and 40 N, making angles of α and β, respectively with the x-direction, act on B as shown in the diagram, and B is moving in the x-direction with constant speed. It is given that $\cos\alpha = 0.6$ and $\cos\beta = 0.8$.

   i Find the total work done by the forces shown in the diagram when B has moved a distance of 20 m. **[2]**

   ii Given that the coefficient of friction between the block and the plane is $\frac{5}{8}$, find the weight of the block. **[3]**

Cambridge International AS & A Level Mathematics 9709 Paper 41 Q2 Nov 2013

A Level questions

Reproduced by permission of Cambridge Assessment International Education

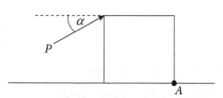

A cube of volume $4096\,cm^3$ and mass $5\,kg$ rests on a rough horizontal surface. The coefficient of friction between the surface and the cube is 0.42. A force P is applied to the top edge of the cube in line with the plane of symmetry of the cube and at an angle α below the horizontal, as shown in the diagram.

Given that $\sin\alpha = \frac{2}{7}$ and that the force P is gradually increased until the cube moves, determine whether the cube will slide or topple first.

2 O, A and B are three points in a straight line on a smooth horizontal surface. A particle P of mass $0.6\,kg$ moves along the line. At time t s the particle has displacement x m from O and speed $v\,m\,s^{-1}$.

The only horizontal force acting on P has magnitude $0.4v^{\frac{1}{2}}$ N and acts in the direction OA. Initially the particle is at A, where $x = 1$ and $v = 1$.

i Show that $3v^{\frac{1}{2}}\dfrac{dv}{dx} = 2$. [2]

ii Express v in terms of x. [4]

iii Given that $AB = 7$ m, find the value of t when P passes through B. [3]

Cambridge International AS & A Level Mathematics 9709 Paper 51 Q6 Nov 2014

3 A small uniform sphere A, of mass $2m$, is moving with speed u on a smooth horizontal surface when it collides directly with a small uniform sphere B, of mass m, which is at rest. The spheres have equal radii and the coefficient of restitution between them is e. Find expressions for the speeds of A and B immediately after the collision. [4]

Subsequently B collides with a vertical wall which is perpendicular to the direction of motion of B. The coefficient of restitution between B and the wall is 0.4. After B has collided with the wall, the speeds of A and B are equal. Find e. [2]

Initially B is at a distance d from the wall. Find the distance of B from the wall when it next collides with A. [4]

Cambridge International AS & A Level Further Mathematics 9231 Paper 21 Q2 Nov 2015

4 A particle P is projected with speed $V \, \mathrm{m\,s^{-1}}$ at an angle of 60° above the horizontal from a point O. At the instant 1s later a particle Q is projected from O with the same initial speed at an angle of 45° above the horizontal. The two particles collide when Q has been in motion for t s.

  i Show that $t = 2.414$, correct to 3 decimal places. [3]

  ii Find the value of V. [4]

The collision occurs after P has passed through the highest point of its trajectory.

  iii Calculate the vertical distance of P below its greatest height when P and Q collide. [4]

Cambridge International AS & A Level Mathematics 9709 Paper 51 Q7 Nov 2015

5

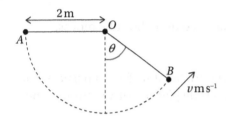

A particle of mass m is attached to one end of a light inextensible string of length 2 m and the other end is attached to a fixed point O. The particle is held horizontally at A with the string taut. The particle is released from rest and moves in a vertical circle until it reaches the point B. At B, the particle is travelling at speed $v \, \mathrm{m\,s^{-1}}$ and the string makes an angle θ with the downward vertical such that $\tan \theta = \dfrac{5}{12}$.

  i Calculate the exact value of the speed, $v \, \mathrm{m\,s^{-1}}$, of the particle at B.

At B, the string breaks and the particle moves as a projectile in the plane OAB. Given that O is 3 m above the ground:

  ii calculate the exact value of the speed at which the particle hits the ground

  iii taking B as the origin, show that the projectile motion is described by the Cartesian equation $y = \dfrac{x}{1152}(480 - 2197x)$.

6 One end of a light elastic string of natural length 1.6 m and modulus of elasticity 28 N is attached to a fixed point O. The other end of the string is attached to a particle P of mass 0.35 kg which hangs in equilibrium vertically below O. The particle P is projected vertically upwards from the equilibrium position with speed 1.8 m s^{-1}. Calculate the speed of P at the instant the string first becomes slack. [5]

Cambridge International AS & A Level Mathematics 9709 Paper 51 Q3 Nov 2014

7

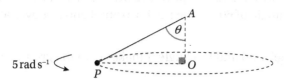

One end of a light inextensible string is attached to a fixed point A and the other end of the string is attached to a particle P. The particle P moves with constant angular speed $5\,\text{rad s}^{-1}$ in a horizontal circle which has its centre O vertically below A. The string makes an angle θ with the vertical (see diagram). The tension in the string is three times the weight of P.

i Show that the length of the string is $1.2\,\text{m}$. [3]

ii Find the speed of P. [4]

Cambridge International AS & A Level Mathematics 9709 Paper 51 Q3 June 2015

8

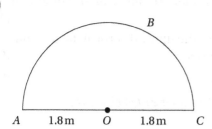

A uniform metal frame $OABC$ is made from a semicircular arc ABC of radius $1.8\,\text{m}$, and a straight rod AOC with $AO = OC = 1.8\,\text{m}$ (see diagram).

i Calculate the distance of the centre of mass of the frame from O. [3]

A uniform semicircular lamina of radius $1.8\,\text{m}$ has weight $27.5\,\text{N}$. A non-uniform object is formed by attaching the frame $OABC$ around the perimeter of the lamina. The object is freely suspended from a fixed point at A and hangs in equilibrium. The diameter AOC of the object makes an angle of $22°$ with the vertical.

ii Calculate the weight of the frame. [5]

Cambridge International AS & A Level Mathematics 9709 Paper 51 Q5 June 2014

9 A uniform ladder AB of mass $8\,\text{kg}$ and length $4\,\text{m}$ rests in equilibrium against a smooth vertical wall and on rough horizontal ground. The ladder meets the wall at a height of $3\,\text{m}$ above the ground. A tin of paint of mass $2\,\text{kg}$ is balanced on one of the steps $\frac{3}{4}$ of the way up the ladder.

Show that the minimum value of the coefficient of friction between the ladder and the ground to prevent the ladder from slipping is $\mu = \dfrac{11\sqrt{7}}{60}$.

10 Joanna throws a ball of mass 0.3 kg underarm with an initial speed of $2\,\text{m s}^{-1}$. The ball leaves Joanna's hands 0.5 m above the ground at an angle of 60° above the horizontal and it moves as a projectile.

   i Find the speed at which the ball hits the ground and the angle it makes with the ground at the point of impact.

   ii On impact with the ground, the ball bounces. The coefficient of restitution between the ball and the ground is 0.4. Find the maximum height of the ball after its first bounce and the angle the ball makes with the ground immediately after the bounce.

11 A particle P of mass m is attached to one end of a light inextensible string of length a. The other end of the string is attached to a fixed point O. When P is hanging at rest vertically below O, it is projected horizontally. In the subsequent motion P completes a vertical circle. The speed of P when it is at its highest point is u. Show that the least possible value of u is $\sqrt{(ag)}$.   **[2]**

  It is now given that $u = \sqrt{(ag)}$. When P passes through the lowest point of its path, it collides with, and coalesces with, a stationary particle of mass $\frac{1}{4}m$. Find the speed of the combined particle immediately after the collision.   **[4]**

  In the subsequent motion, when OP makes an angle θ with the upward vertical the tension in the string is T. Find an expression for T in terms of m, g and θ.   **[5]**

  Find the value of $\cos\theta$ when the string becomes slack.   **[2]**

Cambridge International AS & A Level Further Mathematics 9231 Paper 21 Q4 Nov 2015

12 A small ball B is projected from a point O with speed $14\,\text{m s}^{-1}$ at an angle of 60° above the horizontal.

   i Calculate the speed and direction of motion of B for the instant 1.8 s after projection.   **[5]**

  The point O is 2 m above a horizontal plane.

   ii Calculate the time after projection when B reaches the plane.   **[3]**

Cambridge International AS & A Level Mathematics 9709 Paper 51 Q4 Nov 2013

13 A non-uniform rod AB of weight 6 N rests in limiting equilibrium with the end A in contact with a rough vertical wall. $AB = 1.2\,\text{m}$, the centre of mass of the rod is 0.8 m from A, and the angle between AB and the downward vertical is $\theta°$. A force of magnitude 10 N acting at an angle of 30° to the upwards vertical is applied to the rod at B (see diagram). The rod and the line of action of the 10 N force lie in a vertical plane perpendicular to the wall. Calculate

   i the value of θ **[4]**

   ii the coefficient of friction between the rod and the wall.   **[2]**

Cambridge International AS & A Level Mathematics 9709 Paper 51 Q2 June 2014

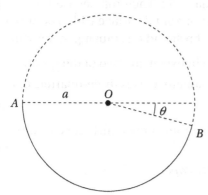

A smooth wire is in the form of an arc AB of a circle, of radius a, that subtends an obtuse angle $\pi - \theta$ at the centre O of the circle. It is given that $\sin \theta = \frac{1}{4}$. The wire is fixed in a vertical plane, with AO horizontal and B below the level of O (see diagram). A small bead of mass m is threaded on the wire and projected vertically downwards from A with speed $\sqrt{\left(\frac{3}{10}ga\right)}$.

i   Find the reaction between the bead and the wire when the bead is vertically below O.   **[3]**

ii Find the speed of the bead as it leaves the wire at B.   **[3]**

iii Show that the greatest height reached by the bead is $\frac{1}{8}a$ above the level of O.   **[4]**

Cambridge International AS & A Level Further Mathematics 9231 Paper 21 Q4 June 2014

15 A light elastic string has natural length 3 m and modulus of elasticity 45 N. A particle P of weight 6 N is attached to the mid-point of the string. The ends of the string are attached to fixed points A and B which lie in the same vertical line with A above B and $AB = 4$ m. The particle P is released from rest at the point 1.5 m vertically below A.

i   Calculate the distance P moves after its release before first coming to instantaneous rest at a point vertically above B. (You may assume that at this point the part of the string joining P to B is slack.)   **[4]**

ii Show that the greatest speed of P occurs when it is 2.1 m below A, and calculate this greatest speed.   **[5]**

iii Calculate the greatest magnitude of the acceleration of P.   **[3]**

Cambridge International AS & A Level Mathematics 9709 Paper 51 Q7 Nov 2012

16 A particle P of mass 0.6 kg is released from rest at a point above ground level and falls vertically. The motion of P is opposed by a force of magnitude $3v$ N, where v m s^{-1} is the speed of P. Immediately before P reaches the ground, $v = 1.95$.

i   Calculate the time after its release when P reaches the ground.   **[5]**

P is now projected horizontally with speed 1.95 m s^{-1} across a smooth horizontal surface. The motion of P is again opposed by a force of magnitude $3v$ N, where v m s^{-1} is the speed of P.

ii Calculate the distance P travels after projection before coming to rest.   **[3]**

Cambridge International AS & A Level Mathematics 9709 Paper 51 Q6 June 2014

17 Three uniform small smooth spheres, A, B and C, have equal radii. Their masses are $4m$, $2m$ and m respectively. They lie in a straight line on a smooth horizontal surface with B between A and C. Initially A is moving towards B with speed u, B is at rest and C is moving in the same direction as A with speed $\frac{1}{2}u$. The coefficient of restitution between any two of the spheres is e. The first collision is between A and B. In this collision sphere A loses three-quarters of its kinetic energy. Show that $e = \frac{1}{2}$. **[6]**

Find the speed of B after its collision with C and deduce that there are no further collisions between the spheres. **[5]**

Cambridge International AS & A Level Further Mathematics 9231 Paper 21 Q2 June 2013

18 A small ball B is projected with speed $15\,\text{m s}^{-1}$ at an angle of $41°$ above the horizontal from a point O which is 1.6 m above horizontal ground. At time t s after projection the horizontal and vertically upward displacements of B from O are x m and y m respectively.

i Express x and y in terms of t and hence show that the equation of the trajectory of B is
$$y = 0.869x - 0.0390x^2,$$
where the coefficients are correct to 3 significant figures. **[4]**

A vertical fence is 1.5 m from O and perpendicular to the plane in which B moves. B just passes over the fence and subsequently strikes the ground at the point A.

ii Calculate the height of the fence, and the distance from the fence to A. **[5]**

Cambridge International AS & A Level Mathematics 9709 Paper 51 Q7 June 2012

19

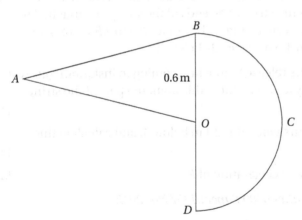

A uniform lamina OABCD consists of a semicircle BCD with centre O and radius 0.6 m and an isosceles triangle OAB, joined along OB (see diagram). The triangle has area $0.36\,\text{m}^2$ and $AB = AO$.

i Show that the centre of mass of the lamina lies on OB. **[4]**

ii Calculate the distance of the centre of mass of the lamina from O. **[4]**

Cambridge International AS & A Level Mathematics 9709 Paper 51 Q6 Nov 2012

20 A particle P of mass m is projected horizontally with speed $\sqrt{\left(\frac{7}{2}ga\right)}$ from the lowest point of the inside of a fixed hollow smooth sphere of internal radius a and centre O. The angle between OP and the downward vertical at O is denoted by θ. Show that, as long as P remains in contact with the inner surface of the sphere, the magnitude of the reaction between the sphere and the particle is $\frac{3}{2}mg(1 + 2\cos\theta)$. [4]

Find the speed of P

 i when it loses contact with the sphere, [3]

 ii when, in the subsequent motion, it passes through the horizontal plane containing O. (You may assume that this happens before P comes into contact with the sphere again.) [3]

Cambridge International AS & A Level Further Mathematics 9231 Paper 21 Q3 June 2012

21 A particle P of mass $0.4\,\text{kg}$ is attached to one end of a light elastic string of natural length $1.2\,\text{m}$ and modulus of elasticity $19.2\,\text{N}$. The other end of the string is attached to a fixed point A. The particle P is released from rest at the point $2.7\,\text{m}$ vertically above A. Calculate

 i the initial acceleration of P, [3]

 ii the speed of P when it reaches A. [4]

Cambridge International AS & A Level Mathematics 9709 Paper 51 Q2 June 2013

22 A particle P of mass $0.8\,\text{kg}$ moves along the x-axis on a horizontal surface. When the displacement of P from the origin O is x m the velocity of P is $v\,\text{m s}^{-1}$ in the positive x-direction. Two horizontal forces act on P. One force has magnitude $4e^{-x}\,\text{N}$ and acts in the positive x-direction. The other force has magnitude $2.4x^2\,\text{N}$ and acts in the negative x-direction.

 i Show that $v\dfrac{dv}{dx} = 5e^{-x} - 3x^2$. [2]

 ii The velocity of P as it passes through O is $6\,\text{m s}^{-1}$. Find the velocity of P when $x = 2$. [5]

Cambridge International AS & A Level Mathematics 9709 Paper 51 Q3 Nov 2013

23 Two uniform small smooth spheres A and B, of equal radii, have masses $2m$ and m respectively. They lie at rest on a smooth horizontal plane. Sphere A is projected directly towards B with speed u. After the collision B goes on to collide directly with a fixed smooth vertical barrier, before colliding with A again. The coefficient of restitution between A and B is $\frac{2}{3}$ and the coefficient of restitution between B and the barrier is e. After the second collision between A and B, the speed of B is five times the speed of A. Find the two possible values of e. [11]

Cambridge International AS & A Level Further Mathematics 9231 Paper 21 Q5 Nov 2013

24 A small ball is projected with speed $20\,\text{m s}^{-1}$ at an angle of $45°$ above the horizontal from a point O on horizontal ground. At time t s after projection, the horizontal and vertically upwards displacements of the ball from O are x m and y m respectively.

 i Express x and y in terms of t. [2]

 ii Show that the equation of the trajectory of the ball is $y = x - \dfrac{1}{40}x^2$. [2]

 iii State the distance from O of the point at which the ball first strikes the ground. [1]

Cambridge International AS & A Level Mathematics 9709 Paper 51 Q1 June 2013

25

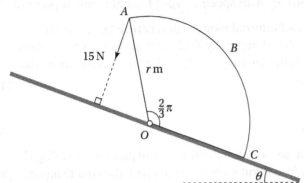

OABC is the cross-section through the centre of mass of a uniform prism of weight 20 N. The cross-section is in the shape of a sector of a circle with centre *O*, radius *OA* = *r* m and angle $AOC = \frac{2}{3}\pi$ radians. The prism lies on a plane inclined at an angle θ radians to the horizontal, where $\theta < \frac{1}{3}\pi$. *OC* lies along a line of greatest slope with *O* higher than *C*. The prism is freely hinged to the plane at *O*. A force of magnitude 15 N acts at *A*, in a direction towards to the plane and at right angles to it (see diagram). Given that the prism remains in equilibrium, find the set of possible values of θ. **[9]**

Cambridge International AS & A Level Mathematics 9709 Paper 51 Q7 June 2013

26 A particle *P* of mass *m* is attached to one end of a light inextensible string of length *a*. The other end of the string is attached to a fixed point *O*. The particle is held with the string taut and horizontal and is then released. When the string is vertical, it comes into contact with a small smooth peg *A* which is vertically below *O* and at a distance *x* (< *a*) from *O*. In the subsequent motion, when *AP* makes an angle θ with the downward vertical, the tension in the string is *T*. Show that

$$T = mg\left(3\cos\theta + \frac{2x}{a - x}\right)$$ **[7]**

Given that *P* completes a vertical circle about *A*, find the least possible value of $\frac{x}{a}$. **[2]**

Cambridge International AS & A Level Further Mathematics 9231 Paper 21 Q3 Nov 2012

27 A light elastic string has natural length 0.8 m and modulus of elasticity 16 N. One end of the string is attached to a fixed point *O*, and a particle *P* of mass 0.4 kg is attached to the other end of the string. The particle *P* hangs in equilibrium vertically below *O*.

i Show that the extension of the string is 0.2 m. **[2]**

P is projected vertically downwards from the equilibrium position. *P* first comes to instantaneous rest at the point where *OP* = 1.4 m.

ii Calculate the speed at which *P* is projected. **[3]**

iii Find the speed of *P* at the first instant when the string subsequently becomes slack. **[2]**

Cambridge International AS & A Level Mathematics 9709 Paper 51 Q3 June 2014

28 A particle P of mass $0.2\,\mathrm{kg}$ is released from rest and falls vertically. At time $t\,\mathrm{s}$ after release P has speed $v\,\mathrm{m\,s^{-1}}$. A resisting force of magnitude $0.8v\,\mathrm{N}$ acts on P.

  i   Show that the acceleration of P is $(10 - 4v)\,\mathrm{m\,s^{-2}}$.   [2]

  ii Find the value of v when $t = 0.6$.   [5]

Cambridge International AS & A Level Mathematics 9709 Paper 51 Q3 Nov 2012

29 Two smooth spheres A and B, of equal radii and masses $2m$ and m respectively, lie at rest on a smooth horizontal table. The spheres A and B are projected directly towards each other with speeds $4u$ and $3u$, respectively. The coefficient of restitution between the spheres is e. Find the set of values of e for which the direction of motion of A is reversed in the collision.   [5]

Cambridge International AS & A Level Further Mathematics 9231 Paper 21 Q1 Nov 2014

Extension questions

1

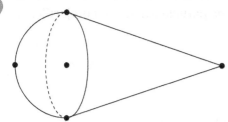

A solid hemisphere and a solid cone are attached by a common vertical circle with radius $10\,\mathrm{cm}$. The density of the shape is uniform and it is perfectly balanced on the circumference of the common circle. The shape is then spun around with angular speed $4\,\mathrm{rad\,s^{-1}}$ such that the vertical diameter of the common circle is the centre of rotation. Calculate the speed at each end of the shape.

2

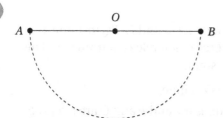

Particle A of mass $m\,\mathrm{kg}$ is attached to the end of a light rod of length $l\,\mathrm{m}$. The other end is attached to a fixed point O. Particle B of mass $3m\,\mathrm{kg}$ is attached to the end of a different light rod of length $l\,\mathrm{m}$. The other end is also attached to O. A and B are held in position such that AOB is a horizontal straight line with O between A and B, as shown in the diagram. A and B are released from rest and they move downwards in a circular arc before colliding. The coefficient of restitution between A and B is 0.4.

  i   Find the velocities of A and B immediately before the collision.

  ii Find the velocities of A and B immediately after the collision.

  iii Find the angle that OA and OB make with the downwards vertical after the collision when A and B come instantaneously to rest.

3 A bouncy ball is dropped vertically downwards from rest at a height of 20 m above the ground. The ball bounces vertically upwards before travelling downwards and bouncing again. The coefficient of restitution between the ball and the ground is 0.85. How many bounces are needed before the ball remains below 5 m?

4 A particle P of mass 2 kg is moving along a straight horizontal line from a fixed point O. After t seconds, the displacement of P from O is x m and its velocity is v ms^{-1}. The resultant force acting on P is $\dfrac{x^2}{\sin\left(\dfrac{v}{2}\right)}$ N in the direction of motion, where $0 < v < 2\pi$. At O, the particle is travelling at π ms^{-1}. Find an expression for x in terms of v.

5 A particle P of mass 0.1 kg is attached to one end of a spring of natural length 0.5 m and modulus of elasticity 10 N. The other end is attached to a fixed point O. The spring is compressed to half its natural length and positioned at an angle of 45° above the horizontal. The spring is released from rest and extends in a straight line in the direction of the spring. At the point when the particle reaches its maximum speed it separates from the spring and moves as a projectile. Find an equation for the path of the particle when it is a projectile, in relation to the fixed point O.

6

$$\begin{array}{cc} \xleftarrow{\hspace{1cm}} 2\,\text{m} \xrightarrow{\hspace{1cm}} & \xleftarrow{\hspace{1cm}} 2.2\,\text{m} \xrightarrow{\hspace{1cm}} \\ P\,(4\,\text{kg}) \qquad\qquad A & B \qquad\qquad Q\,(3\,\text{kg}) \end{array}$$

A particle P of mass 4 kg is attached to one end of a light elastic string of natural length 1.5 m and modulus of elasticity 2 N. The other end is attached to a fixed point A. Particle Q of mass 3 kg is attached to one end of a second light elastic string of natural length 1.2 m and modulus of elasticity 2 N. The other end is attached to a fixed point B. A, B, P and Q lie on a smooth horizontal surface.

Initially P and Q are held in position with $AP = 2$ m and $BQ = 2.2$ m and $PABQ$ forming a horizontal straight line, as shown in the diagram. P and Q are then released from rest. The distance between A and B is l m, where l is a positive constant.

i Write down the range of possible values for l if P and Q collide.

Consider the case where $l = 0$, the coefficient of restitution is 0.7 and P and Q are released such that they collide where A and B now meet.

ii Find the length of BQ after the collision when Q first comes to instantaneous rest.

7 A particle P of mass 4 kg is moving along a straight horizontal line from a fixed point O. After t seconds, the displacement of P from A is x m and its velocity is v ms^{-1}. The resultant force acting on P is $e^{2x}\sin(2x)$ N in the direction of motion. At O, the particle is travelling at 2 ms^{-1}. Find an expression for v in terms of x.

8

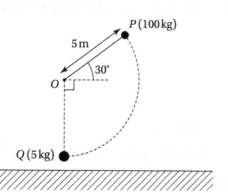

A smooth particle P of mass $100\,\text{kg}$ is attached to one end of a light rod of length $5\,\text{m}$. The other end is attached to a fixed point O. P is initially held in position such that OP makes an angle of $30°$ with the horizontal. P is projected perpendicular to OP with an initial speed of $u\,\text{ms}^{-1}$. P moves in the downwards arc of a circle until it collides with a smooth stationary particle Q of mass $5\,\text{kg}$. P and Q have the same radii and the collision is parallel to the smooth horizontal surface that Q is resting on. Given that the speed of Q after the collision is $5u$, find an expression for the coefficient of restitution in terms of u.

9 A particle P is projected vertically upwards at a speed of $u\,\text{ms}^{-1}$ from a fixed point O on the ground. After 2 seconds, particle Q is projected vertically upwards with a speed of $2u\,\text{ms}^{-1}$ from O. P and Q collide T seconds after the initial projection of P. Find an expression for T in terms of u.

10 A particle P is projected at a speed of $5\,\text{ms}^{-1}$ at an angle of $60°$ above the horizontal from a fixed point O. After T seconds, particle Q is projected at a speed of $8\,\text{ms}^{-1}$ at an angle of $30°$ to the horizontal from O. Given that P and Q collide, find the value of T.

11

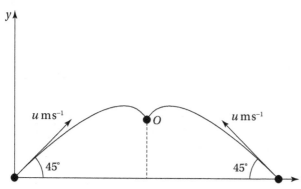

A smooth particle A is projected at $u\,\text{ms}^{-1}$ at an angle of $45°$ above the horizontal from a fixed point O. A smooth particle B is projected at $u\,\text{ms}^{-1}$ at an angle of $45°$ above the horizontal from a fixed point that is $l\,\text{m}$ from O. A and B have the same radii and their motion is symmetrical as shown in the diagram.

i Given that A and B collide, find the range of possible values for l in terms of u.

It is now given that:

➤ A and B have the same mass

➤ $l = 14\,\text{m}$

➤ $u = 10\,\text{m}\,\text{s}^{-1}$

➤ the coefficient of restitution between A and B is 0.5.

ii Find the distance between A and B when they first strike the ground.

12 A particle moves in a straight line and its acceleration, $a\,\text{m}\,\text{s}^{-2}$, varies with its displacement, $x\,\text{m}$, from a fixed origin, O, such that $a = \dfrac{4}{\sqrt{x^2 - 9}}$. When the particle has a displacement of $3\sqrt{65}\,\text{m}$, it has velocity $4\,\text{m}\,\text{s}^{-1}$. Find the possible values of the velocity when the particle has a displacement of $6\,\text{m}$.

GLOSSARY

angular speed As a particle moves around a circle it turns through an angle of θ radians. The speed of the particle is $\frac{s}{t} = \frac{r\theta}{t}$ as $s = r\theta$. The **angular speed**, ω, is defined as $\frac{\theta}{t}$ and is measured in rad s^{-1}.

centre of mass The point of a body where it will balance.

centripetal force The resultant force acting on the particle towards the centre of the circle.

compression The distance that a spring is shortened by squeezing.

conical pendulum A pendulum that consists of a particle attached to a light inextensible string, and in a horizontal circle.

conservation of mechanical energy The principle that states that the sum of the potential and kinetic energy of a system remains constant if no energy is lost.

elastic potential energy The energy stored in a string or spring extended from its natural length.

elasticity The measure of how far a string or spring can be stretched from its natural length.

energy The ability to do work.

equilibrium When forces all balance and, as a result, the sum of the forces in any direction is zero.

extension The distance a string or spring is stretched from its natural length.

force An influence that causes an object to accelerate.

gravitational potential energy The energy stored in an object due to its height above the ground.

inelastic A collision where $\frac{\text{speed of separation of particles}}{\text{speed of approach of particles}} = 0$.

instantaneous velocity The velocity of an object in motion at a specific point in time.

kinetic energy The energy a body has as a result of its motion.

lamina A flat object that has negligible thickness.

limiting equilibrium If friction is limiting but an object remains stationary, it is in limiting equilibrium.

mass A measure of how much matter is in an object.

moment A turning moment is generated when a force is applied to an object that is fixed or supported at a point along an axis that does not pass through the point.

momentum The tendency of a moving object to continue moving.

Newton's experimental law A law that relates the speed of two objects after a collision with the speeds before the collision and the properties of the objects.

oblique impact A non-perpendicular collision.

parabola The shape achieved when a cone is cut by a plane that is at the same incline as the side of the cone.

perfectly elastic A collision where $\frac{\text{speed of separation of particles}}{\text{speed of approach of particles}} = 1$.

period The period of a pendulum is the time taken to complete one full circle.

projectile An object propelled through the air such that its subsequent motion takes place in two dimensions rather than one.

range The horizontal distance travelled by a projectile.

sense The sense of a moment describes whether a moment is clockwise or anticlockwise. Anticlockwise moments are positive and clockwise moments are negative.

sliding If an object is on the point of sliding, then $F = \mu R$.

tangential speed The speed of an object moving in a straight line.

tension The force in a string when it is taut.

terminal velocity The velocity at the point where the magnitude of the air resistance and the weight of a falling object are the same, so there is no resultant force acting on the object.

toppling At the instant that an object on a plane starts to rotate, it is described as being on the point of toppling. Its centre of mass is then directly above the lowest point of contact between the object and the plane.

trajectory The graph of the path of a projectile.

work done The work done against a force of F N for a distance of d m is given by the formula Work done $= Fd$.

INDEX